Preface

This book contains the detailed solutions of all the exercises of my book: Tensor Calculus Made Simple. The solutions are ordered according to their arrangement in the original book, i.e. grouped according to the chapters and numbered as in the original book. The solutions are generally very detailed and hence they are supposed to provide some sort of revision for the subject topic. I hope these solutions will be useful to the readers and users of my original book. However, I strongly encourage the readers and users to refer to these solutions only after they solve the exercises, or at least after they make a serious effort, so that consolidation and acquiring the necessary mathematical skills, which are the objectives of these exercises will not be lost or diminished.

Taha Sochi

London, November 2017

Contents

Table of Contents	2
1 Preliminaries	3
2 Tensors	24
3 Tensor Operations	50
4 δ and ϵ Tensors	58
5 Applications of Tensor Notation and Techniques	68
6 Metric Tensor	79
7 Covariant Differentiation	84

Chapter 1
Preliminaries

1.1 Name three mathematicians accredited for the development of tensor calculus. For each one of these mathematicians, give a mathematical technical term that bears his name.
Mathematicians: Ricci, Levi-Civita and Christoffel. **Terms**: Ricci calculus, Levi-Civita symbol, Christoffel symbols.

1.2 What are the main scientific disciplines that employ the language and techniques of tensor calculus?
Differential geometry, general theory of relativity, continuum mechanics, and fluid dynamics among other disciplines.

1.3 Mention one cause for the widespread use of tensor calculus in science.
Its ability to deal with complex problems in multi-dimensional spaces as it provides very efficient and powerful mathematical tools for modeling and analyzing this type of problems while complying with the principle of invariance.

1.4 Describe some of the distinctive features of tensor calculus which contributed to its success and extensive use in mathematics, science and engineering.
Elegance, compactness, clarity, power, eloquence, etc.

1.5 Give preliminary definitions of the following terms: scalar, vector, tensor, rank of tensor, and dyad.
Scalar: mathematical object (e.g. number, variable, function) that is completely identified by its magnitude and sign with no reference to a basis vector set.
Vector: mathematical object that is completely identified by its magnitude and a single direction in space; hence it requires a basis vector set to be fully identified in component form.
Tensor: mathematical object that is completely identified by a set of components each of which is associated with multiple directions in space and hence it requires a basis tensor set (e.g. $\mathbf{E}_i \mathbf{E}_j$ for rank-2). Tensor is also used to describe some mathematical objects with certain transformation properties and hence it includes scalars and vectors.

Rank of tensor: a measure of the complexity of the structure of tensor; it is quantified by the number of free indices of the tensor.

Dyad: tensor object associated with double direction; outer product of two vectors such as $\mathbf{E}_1 \mathbf{E}_3$.

1.6 What is the meaning of the following mathematical symbols?

$$\partial_i \qquad \partial_{ii} \qquad \nabla \qquad A_{,i} \qquad \Delta \qquad A_{i;k}$$

Partial derivative operator with respect to the ith coordinate.
Laplacian operator in Cartesian form.
nabla operator.
Partial derivative of scalar A with respect to the ith coordinate.
Laplacian operator.
Covariant derivative of vector A_i (which is a rank-1 covariant tensor) with respect to the kth coordinate.

1.7 Is the following equality correct? If so, is there any condition for this to hold?

$$\partial_k \partial_l = \partial_l \partial_k$$

Yes, it is correct assuming the C^2 continuity condition.

1.8 Pick five terms from the Index about partial derivative operators and their symbols as used in tensor calculus and explain each one of these terms giving some examples for their use.
Comma: used as subscript to indicate partial differentiation with respect to the following index(es). Examples: $f_{,i} \quad A^i{}_{,jk} \quad B_{ij,k}$.
Commutative: partial differential operators with respect to different variables are commutative assuming the C^2 continuity condition. Examples: $\partial_i \partial_j g = \partial_j \partial_i g$, $\partial_k \partial_l C_q = \partial_l \partial_k C_q$.
Curl: cross product of the nabla operator and a non-scalar tensor. Examples: $\nabla \times \mathbf{A}$, $\nabla \times \mathbf{B}$.
Laplacian: operator representing consecutive application of spatial partial differentiation as represented by the nabla operator. It may be described as the divergence of gradient. It is usually symbolized by ∇^2 or Δ or ∂_{ii}. Examples: $\nabla^2 f$, $\Delta \mathbf{B}$, $\partial_{jj} A_i$.
nabla: partial differential operator usually symbolized by ∇. It is vector operator. It

is spatial operator as it stands for differentiation with respect to coordinates of space. It is used as a basis for a number of operators and operations in vector and tensor calculus such as gradient, divergence and curl. Examples: ∇f, $\nabla \cdot \mathbf{A}$.

1.9 Describe, briefly, the following six coordinate systems outlining their main features: orthonormal Cartesian, cylindrical, plane polar, spherical, general curvilinear, and general orthogonal.

Orthonormal Cartesian: rectilinear coordinate system with mutually orthogonal basis vectors each of unit length; usually represented in 3D space by (x, y, z) where each one of these variables represents signed distance from the origin of coordinates along the corresponding coordinate axis; the simplest and most commonly used of all coordinate systems; should be chosen if it satisfies the requirements for solving the given problem.

Plane polar: it is the same as the cylindrical system (refer to the next) with the absence of the third coordinate (i.e. z coordinate); hence it is 2D system used in coordinating plane surfaces; particularly useful for solving problems with circular symmetry and problems of rotational nature.

Cylindrical: curvilinear orthogonal coordinate system; it has unit basis vectors by definition; these vectors are coordinate dependent in their direction except the basis vector corresponding to the third coordinate which is constant; specifically defined for 3D spaces; particularly useful for solving problems with cylindrical symmetry; usually represented by (ρ, ϕ, z) where ρ represents the perpendicular positive distance from the given point in space to the third axis of the corresponding rectangular Cartesian system, ϕ represents the angle between the first axis and the line connecting the origin of coordinates to the perpendicular projection of the point on the $x_1 x_2$ plane of the corresponding Cartesian system, and z is the same as the third coordinate of the point in the reference Cartesian system.

Spherical: curvilinear orthogonal coordinate system; it has unit basis vectors by definition; all these vectors are coordinate dependent in their direction; specifically defined for 3D spaces; particularly useful for solving problems with spherical symmetry; usually represented by (r, θ, ϕ) where r represents the positive distance from the origin of coordinates to the given point in space, θ is the angle from the positive x_3 axis of the corresponding Cartesian system to the line connecting the origin of coordinates to the point, and ϕ is as defined in the cylindrical system.

General curvilinear: it includes all types of curvilinear systems; its basis vectors are generally functions of position in space and hence they are coordinate dependent; its basis vectors are not necessarily of unit length or mutually perpendicular; its basis vectors can be covariant of tangential nature or contravariant of gradient nature.

General orthogonal: it can be rectilinear or curvilinear; its characteristic feature is that its basis vectors (whether covariant or contravariant) are mutually perpendicular at each point of the space where the system is defined.

1.10 Which of the six coordinate systems in the previous exercise are orthogonal?
All, except general curvilinear which can be orthogonal or not.

1.11 What "basis vectors" of a coordinate system means and what purpose they serve?
Basis vectors are a set of rank-1 tensors that associates a given coordinate system where non-scalar tensors are identified in reference to that set. Basis vectors can be either of covariant type representing tangents to coordinate curves (i.e. curves of constant coordinates except one) or of contravariant type representing gradients of coordinate surfaces (i.e. surfaces of variable coordinates except one). A coordinate system in nD space requires exactly n mutually independent basis vectors to be fully and unambiguously represented. The main purpose of the basis vectors is to identify non-scalar tensors in component form. Basis vectors are also used in the definition and formulation of other mathematical objects such as metric tensor.

1.12 Which of the six coordinate systems mentioned in the previous exercises have constant basis vectors (i.e. some or all of their basis vectors are constant both in magnitude and in direction)?
Orthonormal Cartesian whose all basis vectors are constant in magnitude and direction.
Cylindrical whose z basis vector is constant in magnitude and direction.

1.13 Which of the above six coordinate systems have unit basis vectors by definition or convention?
All except general curvilinear and general orthogonal.

1.14 Explain the meaning of the coordinates in the cylindrical and spherical systems (i.e. ρ, ϕ and z for the cylindrical, and r, θ and ϕ for the spherical).
Cylindrical: ρ represents the perpendicular positive distance from the given point in space to the third axis of the corresponding rectangular Cartesian system, ϕ represents the angle between the first axis of the corresponding Cartesian system and the line

connecting the origin of coordinates to the perpendicular projection of the point on the $x_1 x_2$ plane of the corresponding Cartesian system, and z is the same as the third coordinate of the point in the reference Cartesian system.

Spherical: r represents the positive distance from the origin of coordinates to the given point in space, θ is the angle from the positive x_3 axis of the corresponding Cartesian system to the line connecting the origin of coordinates to the point, and ϕ is as defined in the cylindrical system.

1.15 What is the relation between the cylindrical and plane polar coordinate systems?

Plane polar coordinate system is a cylindrical system with no z dimension and hence it is a 2D system with ρ and ϕ coordinates only.

1.16 Is there any common coordinates between the above six coordinate systems? If so, what? Investigate this thoroughly by comparing each pair of these systems.

Cartesian and cylindrical have common z coordinate. Plane polar and cylindrical have common ρ and ϕ coordinates. Plane polar and spherical have common ϕ coordinate. Cylindrical and spherical have common ϕ coordinate. No common coordinate between the other pairs although there may be accidentally.

1.17 Write the transformation equations between the following coordinate systems in both directions: Cartesian and cylindrical, and Cartesian and spherical.

Cartesian to cylindrical:

$$\rho = \sqrt{x^2 + y^2} \qquad \phi = \arctan\left(\frac{y}{x}\right) \qquad z = z$$

Cylindrical to Cartesian:

$$x = \rho \cos \phi \qquad y = \rho \sin \phi \qquad z = z$$

Cartesian to spherical:

$$r = \sqrt{x^2 + y^2 + z^2} \qquad \theta = \arccos\left(\frac{z}{\sqrt{x^2 + y^2 + z^2}}\right) \qquad \phi = \arctan\left(\frac{y}{x}\right)$$

Spherical to Cartesian:

$$x = r \sin \theta \cos \phi \qquad y = r \sin \theta \sin \phi \qquad z = r \cos \theta$$

1.18 Make a sketch representing a spherical coordinate system, with its basis vectors, superimposed on a rectangular Cartesian system in a standard position.

A sketch should look like Figure 1.

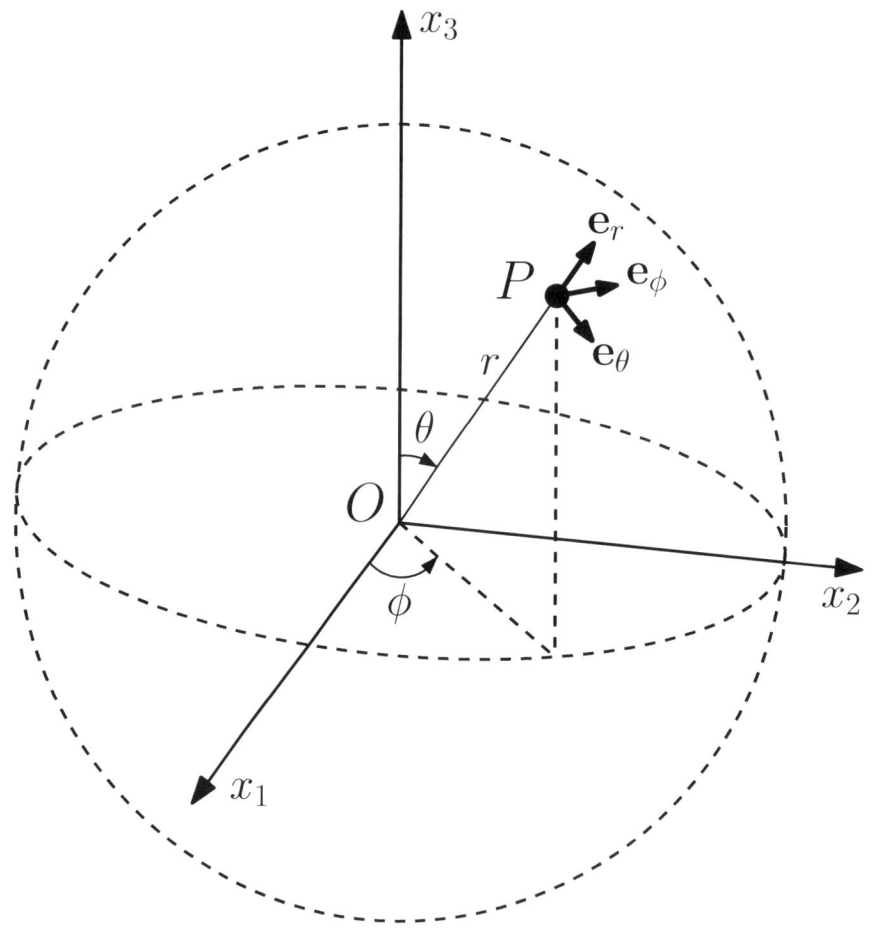

Figure 1: Spherical coordinate system.

1.19 What are the geometric and algebraic definitions of the dot product of two vectors? What is the interpretation of the geometric definition?

Geometric definition: dot product of two vectors is the signed projection of one vector onto the other vector times the length of the other vector, that is:

$$\mathbf{a} \cdot \mathbf{b} = |\mathbf{a}| \, |\mathbf{b}| \cos \theta$$

It can also be defined as the product of the lengths of the two vectors times the cosine of the angle between them. Both these forms of the geometric definition are represented by the above equation although the latter may be easier to grab from the equation.

Algebraic definition: dot product of two vectors is the sum of the products of the corresponding components of the two vectors, that is:

$$\mathbf{a} \cdot \mathbf{b} = a_i b_i \qquad \text{(with summation over } i\text{)}$$

Interpretation of geometric definition: as given above by either of the two forms assuming that the definition is stated by the other form.

1.20 What are the geometric and algebraic definitions of the cross product of two vectors? What is the interpretation of the geometric definition?

Geometric definition: a vector whose length is equal to the area of the parallelogram defined by the two vectors as its two main sides when their tails coincide and whose orientation is perpendicular to the plane of the parallelogram with a direction defined by the right hand rule, that is:

$$\mathbf{a} \times \mathbf{b} = (|\mathbf{a}|\,|\mathbf{b}|\sin\theta)\,\mathbf{n}$$

where $0 \leq \theta \leq \pi$ is the angle between the two vectors when their tails coincide and $\mathbf{n}$ is a unit vector perpendicular to the plane containing $\mathbf{a}$ and $\mathbf{b}$ and is directed according to the right hand rule.

Algebraic definition: a vector represented by the following determinantal form where the determinant (which is not really determinant in strict technical sense) is expanded along its first row, that is:

$$\mathbf{a} \times \mathbf{b} = \begin{vmatrix} \mathbf{e}_1 & \mathbf{e}_2 & \mathbf{e}_3 \\ a_1 & a_2 & a_3 \\ b_1 & b_2 & b_3 \end{vmatrix} = (a_2 b_3 - a_3 b_2)\,\mathbf{e}_1 + (a_3 b_1 - a_1 b_3)\,\mathbf{e}_2 + (a_1 b_2 - a_2 b_1)\,\mathbf{e}_3$$

Interpretation of geometric definition: as stated in the geometric definition prior to the equation.

1.21 What is the dot product of the vectors $\mathbf{A}$ and $\mathbf{B}$ if $\mathbf{A} = (1.9, -6.3, 0)$ and $\mathbf{B} = (-4, -0.34, 11.9)$?

$$\mathbf{A} \cdot \mathbf{B} = (1.9)(-4) + (-6.3)(-0.34) + (0)(11.9) = -5.458$$

1.22 What is the cross product of the vectors in the previous exercise? Write this cross product in its determinantal form and expand it.

$$
\begin{aligned}
\mathbf{A} \times \mathbf{B} &= \begin{vmatrix} \mathbf{e}_1 & \mathbf{e}_2 & \mathbf{e}_3 \\ 1.9 & -6.3 & 0 \\ -4 & -0.34 & 11.9 \end{vmatrix} \\
&= [(-6.3)(11.9) - (0)(-0.34)]\,\mathbf{e}_1 + \\
&\quad [(0)(-4) - (1.9)(11.9)]\,\mathbf{e}_2 + \\
&\quad [(1.9)(-0.34) - (-6.3)(-4)]\,\mathbf{e}_3 \\
&= -74.97\mathbf{e}_1 - 22.61\mathbf{e}_2 - 25.846\mathbf{e}_3
\end{aligned}
$$

1.23 Define the scalar triple product operation of three vectors geometrically and algebraically.
Geometric definition: the scalar triple product of three vectors $\mathbf{a}, \mathbf{b}$ and $\mathbf{c}$ is a scalar defined by:
$$\mathbf{a} \cdot (\mathbf{b} \times \mathbf{c}) = |\mathbf{a}|\,|\mathbf{b}|\,|\mathbf{c}|\sin\phi\cos\theta$$
where ϕ is the angle between $\mathbf{b}$ and $\mathbf{c}$ while θ is the angle between $\mathbf{a}$ and $\mathbf{b} \times \mathbf{c}$.
Algebraic definition: the scalar triple product is defined as the determinant of the array formed by the components of the three vectors as its rows or columns in the given order, that is:

$$
\mathbf{a} \cdot (\mathbf{b} \times \mathbf{c}) = \det(\mathbf{a},\mathbf{b},\mathbf{c}) = \begin{vmatrix} a_1 & a_2 & a_3 \\ b_1 & b_2 & b_3 \\ c_1 & c_2 & c_3 \end{vmatrix} = \begin{vmatrix} a_1 & b_1 & c_1 \\ a_2 & b_2 & c_2 \\ a_3 & b_3 & c_3 \end{vmatrix}
$$

1.24 What is the geometric interpretation of the scalar triple product? What is the condition for this product to be zero?
The geometric interpretation of the scalar triple product is that its magnitude is equal to the volume of the parallelepiped whose three main sides are the three vectors while its sign is positive or negative depending on whether the vectors form a right-handed or left-handed system. It is zero when the three vectors are coplanar (including collinear) since the volume is zero in this case.

1.25 Is it necessary to use parentheses in the writing of scalar triple products and why? Is it possible to interchange the dot and cross symbols in the product?

1 PRELIMINARIES

It is not necessary to use parentheses since cross product with scalar is not defined and hence $\mathbf{a} \cdot \mathbf{b} \times \mathbf{c}$ for example necessarily means $\mathbf{a} \cdot (\mathbf{b} \times \mathbf{c})$ since $(\mathbf{a} \cdot \mathbf{b}) \times \mathbf{c}$ is meaningless; however parenthesis may be used for more clarity especially in introductory texts where the readers may not be sufficiently familiar with the notations and rules as well as similar reasons.

Yes, it is possible to interchange the dot and cross symbols, hence $\mathbf{a} \cdot (\mathbf{b} \times \mathbf{c}) = (\mathbf{a} \times \mathbf{b}) \cdot \mathbf{c}$.

1.26 Calculate the following scalar triple products:

$$\mathbf{a} \cdot (\mathbf{b} \times \mathbf{c}) \qquad \mathbf{a} \cdot (\mathbf{d} \times \mathbf{c}) \qquad \mathbf{d} \cdot (\mathbf{c} \times \mathbf{b}) \qquad \mathbf{a} \cdot (\mathbf{c} \times \mathbf{b}) \qquad (\mathbf{a} \times \mathbf{b}) \cdot \mathbf{c}$$

where $\mathbf{a} = (7, -0.4, 9.5)$, $\mathbf{b} = (-12.9, -11.7, 3.1)$, $\mathbf{c} = (2.4, 22.7, -6.9)$ and $\mathbf{d} = (-56.4, 29.5, 33.8)$. Note that some of these products may be found directly from other products with no need for detailed calculations.

$$\mathbf{a} \cdot (\mathbf{b} \times \mathbf{c}) = \det(\mathbf{a}, \mathbf{b}, \mathbf{c}) = \begin{vmatrix} 7 & -0.4 & 9.5 \\ -12.9 & -11.7 & 3.1 \\ 2.4 & 22.7 & -6.9 \end{vmatrix}$$
$$= 7\left[(-11.7)(-6.9) - (3.1)(22.7)\right]$$
$$\quad -0.4\left[((3.1)(2.4)) - (-12.9)(-6.9)\right]$$
$$\quad +9.5\left[(-12.9)(22.7) - (-11.7)(2.4)\right]$$
$$= -2409.977$$

$$\mathbf{a} \cdot (\mathbf{d} \times \mathbf{c}) = \det(\mathbf{a}, \mathbf{d}, \mathbf{c}) = \begin{vmatrix} 7 & -0.4 & 9.5 \\ -56.4 & 29.5 & 33.8 \\ 2.4 & 22.7 & -6.9 \end{vmatrix} = -19507.714$$

$$\mathbf{d} \cdot (\mathbf{c} \times \mathbf{b}) = \det(\mathbf{d}, \mathbf{c}, \mathbf{b}) = \begin{vmatrix} -56.4 & 29.5 & 33.8 \\ 2.4 & 22.7 & -6.9 \\ -12.9 & -11.7 & 3.1 \end{vmatrix} = 11939.169$$

$$\mathbf{a} \cdot (\mathbf{c} \times \mathbf{b}) = \det(\mathbf{a}, \mathbf{c}, \mathbf{b}) = -\det(\mathbf{a}, \mathbf{b}, \mathbf{c}) = 2409.977$$

$$(\mathbf{a} \times \mathbf{b}) \cdot \mathbf{c} = \mathbf{a} \cdot (\mathbf{b} \times \mathbf{c}) = -2409.977$$

1.27 Write the twelve possibilities of the scalar triple product of three vectors $\mathbf{a}$, $\mathbf{b}$ and $\mathbf{c}$ and divide them into two sets where the entries in each set are equal. What is the relation between the two sets?

1st set: $\mathbf{a} \cdot (\mathbf{b} \times \mathbf{c})$, $\mathbf{c} \cdot (\mathbf{a} \times \mathbf{b})$, $\mathbf{b} \cdot (\mathbf{c} \times \mathbf{a})$, $(\mathbf{a} \times \mathbf{b}) \cdot \mathbf{c}$, $(\mathbf{c} \times \mathbf{a}) \cdot \mathbf{b}$, $(\mathbf{b} \times \mathbf{c}) \cdot \mathbf{a}$.

2nd set: $\mathbf{a} \cdot (\mathbf{c} \times \mathbf{b})$, $\mathbf{c} \cdot (\mathbf{b} \times \mathbf{a})$, $\mathbf{b} \cdot (\mathbf{a} \times \mathbf{c})$, $(\mathbf{b} \times \mathbf{a}) \cdot \mathbf{c}$, $(\mathbf{a} \times \mathbf{c}) \cdot \mathbf{b}$, $(\mathbf{c} \times \mathbf{b}) \cdot \mathbf{a}$.

The two sets are opposite in sign but identical in absolute value.

1.28 What is the vector triple product of three vectors in mathematical terms? Is it scalar or vector? Is it associative?

The vector triple product of three vectors $\mathbf{a}$, $\mathbf{b}$ and $\mathbf{c}$ is defined mathematically by:

$$\mathbf{a} \times (\mathbf{b} \times \mathbf{c}) \qquad \text{or} \qquad (\mathbf{a} \times \mathbf{b}) \times \mathbf{c}$$

It is vector.

It is not associative, that is:

$$\mathbf{a} \times (\mathbf{b} \times \mathbf{c}) \neq (\mathbf{a} \times \mathbf{b}) \times \mathbf{c}$$

1.29 Give the mathematical expression for the nabla ∇ differential operator in Cartesian systems.

Answer:
$$\nabla_i = \frac{\partial}{\partial x_i} = \partial_i \qquad \text{or} \qquad \nabla = \mathbf{e}_i \frac{\partial}{\partial x_i}$$

For 3D Cartesian systems, it is commonly given in vector notation by:

$$\nabla = \mathbf{i}\frac{\partial}{\partial x} + \mathbf{j}\frac{\partial}{\partial y} + \mathbf{k}\frac{\partial}{\partial z}$$

1.30 State the mathematical definition of the gradient of a scalar field f in Cartesian coordinates. Is it scalar or vector?

Answer:
$$\frac{\partial f}{\partial x_i} = \partial_i f \qquad \text{or} \qquad \nabla = \mathbf{e}_i \frac{\partial f}{\partial x_i}$$

For 3D Cartesian systems, it is commonly given in vector notation by:

$$\nabla f = \mathbf{i}\frac{\partial f}{\partial x} + \mathbf{j}\frac{\partial f}{\partial y} + \mathbf{k}\frac{\partial f}{\partial z}$$

It is vector.

1.31 Is the gradient operation commutative, associative or distributive? Express these

1 PRELIMINARIES

properties mathematically.

Not commutative, not associative but distributive, that is:

$$\nabla f \neq f\nabla$$
$$(\nabla f)h \neq \nabla(fh)$$
$$\nabla(f+h) = \nabla f + \nabla h$$

1.32 What is the relation between the gradient of a scalar field f and the surfaces of constant f? Make a simple sketch to illustrate this relation.

The gradient vector is perpendicular to the surfaces of constant f. A sketch should look like Figure 2.

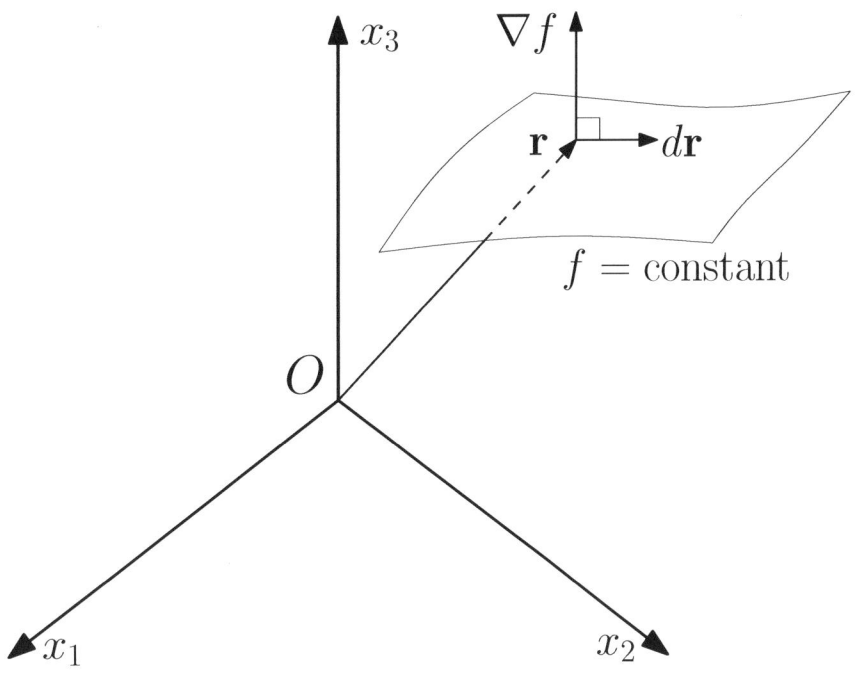

Figure 2: Gradient of scalar field.

1.33 Define, mathematically, the divergence of a vector field $\mathbf{V}$ in Cartesian coordinates. Is it scalar or vector?

The divergence of a vector field $\mathbf{V}$ in a 3D Cartesian system is:

$$\nabla \cdot \mathbf{V} = \frac{\partial V_x}{\partial x} + \frac{\partial V_y}{\partial y} + \frac{\partial V_z}{\partial z}$$

It is scalar.

1.34 Define "solenoidal" vector field descriptively and mathematically.

A vector field **a** is solenoidal *iff* its divergence vanishes identically, that is: $\nabla \cdot \mathbf{a} = 0$ throughout the region of space over which it is defined.

1.35 What is the physical significance of the divergence of a vector field?

It quantifies the convergence or divergence of the field over its domain of definition.

1.36 Is the divergence operation commutative, associative or distributive? Give your answer in words and in mathematical forms.

Not commutative, not associative but distributive, that is:

$$\nabla \cdot \mathbf{A} \neq \mathbf{A} \cdot \nabla$$
$$\nabla \cdot (f\mathbf{A}) \neq \nabla f \cdot \mathbf{A}$$
$$\nabla \cdot (\mathbf{A} + \mathbf{B}) = \nabla \cdot \mathbf{A} + \nabla \cdot \mathbf{B}$$

1.37 Define the curl of a vector field **V** in Cartesian coordinates using the determinantal and the expanded forms with full explanation of all the symbols involved. Is the curl scalar or vector?

Determinantal form:

$$\nabla \times \mathbf{V} = \begin{vmatrix} \mathbf{i} & \mathbf{j} & \mathbf{k} \\ \frac{\partial}{\partial x} & \frac{\partial}{\partial y} & \frac{\partial}{\partial z} \\ V_x & V_y & V_z \end{vmatrix}$$

Expanded form:

$$\nabla \times \mathbf{V} = \left(\frac{\partial V_z}{\partial y} - \frac{\partial V_y}{\partial z}\right)\mathbf{i} + \left(\frac{\partial V_x}{\partial z} - \frac{\partial V_z}{\partial x}\right)\mathbf{j} + \left(\frac{\partial V_y}{\partial x} - \frac{\partial V_x}{\partial y}\right)\mathbf{k}$$

Symbols: ∇ nabla operator. $\times$ cross product operator. $\mathbf{i}, \mathbf{j}, \mathbf{k}$ basis vectors of orthonormal Cartesian 3D system. ∂ partial differentiation symbol. V_x, V_y, V_z components of **V**.

The curl is vector.

1.38 What is the physical significance of the curl of a vector field?

It quantifies the circulation of the field over its domain of definition.

1.39 What is the technical term used to describe a vector field whose curl vanishes identically?

Irrotational.

1 PRELIMINARIES 15

1.40 Is the curl operation commutative, associative or distributive? Express these properties symbolically.
Not commutative, not associative but distributive, that is:

$$\nabla \times \mathbf{A} \neq \mathbf{A} \times \nabla$$
$$\nabla \times (\mathbf{A} \times \mathbf{B}) \neq (\nabla \times \mathbf{A}) \times \mathbf{B}$$
$$\nabla \times (\mathbf{A} + \mathbf{B}) = \nabla \times \mathbf{A} + \nabla \times \mathbf{B}$$

1.41 Describe, in words, the Laplacian operator ∇^2 and how it is obtained. What are the other symbols used to denote it?
It is the divergence of the gradient operator, and hence it is obtained by the following operation:

$$\nabla^2 = \nabla \cdot \nabla$$

where $\nabla \cdot$ means taking the divergence of the gradient operator which is represented by the second ∇.
Examples of other symbols are: $\Delta \quad \partial_{ii} \quad \partial_i \partial_i$.

1.42 Give the mathematical expression of the Laplacian operator in Cartesian systems. Using this mathematical expression, explain why the Laplacian is a scalar rather than a vector operator?
Answer:

$$\nabla^2 = \frac{\partial^2}{\partial x^2} + \frac{\partial^2}{\partial y^2} + \frac{\partial^2}{\partial z^2}$$

As we see, the basis vectors $\mathbf{i}, \mathbf{j}, \mathbf{k}$ are not present in the expression on the right hand side and hence the Laplacian is a scalar operator.

1.43 Can the Laplacian operator act on rank-0, rank-1 and rank-n $(n > 1)$ tensor fields? If so, what is the rank of the resulting field in each case?
Yes, it can.
The Laplacian does not change the rank of the tensor that it acts upon and hence the tensors resulting from applying the Laplacian to the above fields will be of rank-0, rank-1 and rank-n respectively.

1.44 Collect from the Index all the terms related to the nabla based differential operations and classify them according to each one of these operations.
Some of these terms with their classification are:

(a) Gradient, vector operator, contravariant basis vector.
(b) Divergence, solenoidal, divergence theorem, convergence and divergence of field.
(c) Curl, irrotational, Stokes theorem.
(d) Laplacian, scalar operator, harmonic operator.

1.45 Write down the mathematical expression for the divergence theorem, defining all the symbols involved, and explain the meaning of this theorem in words.
Answer:
$$\iiint_V \nabla \cdot \mathbf{A}\, d\tau = \iint_S \mathbf{A} \cdot \mathbf{n}\, d\sigma$$
Symbols: $\mathbf{A}$ is differentiable vector field, V is finite bounded region in space, S is surface enclosing V, $d\tau$ is infinitesimal volume element, $d\sigma$ is infinitesimal area element, $\mathbf{n}$ is unit vector normal to surface, $\iiint$ and $\iint$ are symbols for volume and surface integrals in multi-dimensional spaces, ∇ is nabla operator, $\cdot$ is dot representing dot product.
Meaning: the integral of the divergence of a vector field over a given finite region in space is equal to the total flux of the vector field out of the surface which encloses the region.

1.46 What are the main uses of the divergence theorem in mathematics and science? Explain why this theorem is very useful theoretically and practically.
One of the main uses is converting between volume and surface integrals to simplify calculations. It is also used in mathematical proofs and theoretical arguments.
It is very useful because it simplifies calculations, overcomes analytical and numerical hurdles and provides theoretical foundations.

1.47 If a vector field is given in Cartesian coordinates by $\mathbf{A} = (-0.5, 9.3, 6.5)$, verify the divergence theorem for a cube defined by the plane surfaces $x_1 = -1$, $x_2 = 1$, $y_1 = -1$, $y_2 = 1$, $z_1 = -1$, and $z_2 = 1$.
This is a constant vector field and hence its divergence is zero, that is:
$$\iiint_V \nabla \cdot \mathbf{A}\, d\tau = \iiint_V 0\, d\tau = 0$$

Regarding the surface integral, we see that the integral on the surface $x_1 = -1$ is opposite in sign to the integral on the surface $x_2 = 1$ because the normal vector is opposite in direction on these surfaces while the vector field is the same on both surfaces since it is constant; hence they cancel each other. This also applies to the surfaces y_1, y_2 and z_1, z_2. Consequently the surface integral is also zero, that is:

$$\iint_S \mathbf{A} \cdot \mathbf{n}\, d\sigma = 0$$

So, the divergence theorem is verified in this case.

1.48 Write down the mathematical expression for Stokes theorem with the definition of all the symbols involved and explain its meaning in words. What is this theorem useful for? Why it is very useful?

Answer:

$$\iint_S (\nabla \times \mathbf{A}) \cdot \mathbf{n}\, d\sigma = \int_C \mathbf{A} \cdot d\mathbf{r}$$

Symbols: $\mathbf{A}$ is differentiable vector field, S is open finite surface, C is perimeter of surface S, $d\sigma$ is infinitesimal area element, $d\mathbf{r}$ is infinitesimal tangent vector to perimeter, $\mathbf{n}$ is unit vector normal to surface, $\iint$ and $\int$ are symbols for surface and line integrals, ∇ is nabla operator, $\times$ is cross representing cross product, $\cdot$ is dot representing dot product.

Meaning: the integral of the curl of a differentiable vector field over a finite open surface is equal to the line integral of the vector field around the perimeter surrounding the surface.

Stokes theorem is useful for converting surface integral to line integral and vice versa. It is also useful in mathematical proofs and theoretical arguments.

It can ease the calculations substantially, overcome mathematical hurdles and provide theoretical foundations for many mathematical and physical arguments.

1.49 If a vector field is given in Cartesian coordinates by $\mathbf{A} = (2y, -3x, 1.5z)$, verify Stokes theorem for a hemispherical surface $x^2 + y^2 + z^2 = 9$ for $z \geq 0$.

Left hand side:

$$\begin{aligned}
\nabla \times \mathbf{A} &= \begin{vmatrix} \mathbf{i} & \mathbf{j} & \mathbf{k} \\ \frac{\partial}{\partial x} & \frac{\partial}{\partial y} & \frac{\partial}{\partial z} \\ 2y & -3x & 1.5z \end{vmatrix} \\
&= \left(\frac{\partial (1.5z)}{\partial y} - \frac{\partial (-3x)}{\partial z} \right) \mathbf{i} + \left(\frac{\partial (2y)}{\partial z} - \frac{\partial (1.5z)}{\partial x} \right) \mathbf{j} + \left(\frac{\partial (-3x)}{\partial x} - \frac{\partial (2y)}{\partial y} \right) \mathbf{k} \\
&= 0\mathbf{i} + 0\mathbf{j} - 5\mathbf{k} \\
&= -5\mathbf{k}
\end{aligned}$$

$$\mathbf{n} = \frac{1}{\sqrt{x^2+y^2+z^2}}(x\mathbf{i}+y\mathbf{j}+z\mathbf{k}) = \frac{1}{3}(x\mathbf{i}+y\mathbf{j}+z\mathbf{k})$$

$$(\nabla \times \mathbf{A}) \cdot \mathbf{n} = (-5\mathbf{k}) \cdot \left(\frac{x\mathbf{i}+y\mathbf{j}+z\mathbf{k}}{3}\right) = -\frac{5}{3}z$$

$$\iint_S (\nabla \times \mathbf{A}) \cdot \mathbf{n}\, d\sigma = \iint_S -\frac{5}{3}z\, d\sigma = -\frac{5}{3}\iint_S z\, d\sigma$$

On transforming to spherical coordinates, we have:

$$\begin{aligned}
\iint_S z\, d\sigma &= \int_{\phi=0}^{2\pi}\int_{\theta=0}^{\pi/2} r\cos\theta\; r^2\sin\theta\, d\theta\, d\phi \\
&= r^3 \int_{\phi=0}^{2\pi}\int_{\theta=0}^{\pi/2} \cos\theta\,\sin\theta\, d\theta\, d\phi \\
&= r^3 \int_{\phi=0}^{2\pi} \left[\frac{\sin^2\theta}{2}\right]_0^{\pi/2} d\phi \\
&= r^3 \int_{\phi=0}^{2\pi} \frac{1}{2} d\phi \\
&= \frac{r^3}{2} \int_{\phi=0}^{2\pi} d\phi \\
&= \frac{r^3}{2}(2\pi) \\
&= r^3\pi \\
&= 27\pi
\end{aligned}$$

Hence:

$$\iint_S (\nabla \times \mathbf{A}) \cdot \mathbf{n}\, d\sigma = -\frac{5}{3}\iint_S z\, d\sigma = -45\pi$$

Right hand side:

The perimeter is the circle in the xy plane ($z = 0$) with radius 3 and the sense of traversing the circle is anticlockwise as seen from above, i.e. from positive x axis to positive y axis. By transforming to cylindrical coordinates, we have $x = 3\cos\phi$ and $y = 3\sin\phi$ where $0 \leq \phi < 2\pi$. Hence:

$$\mathbf{r}(\phi) = x\mathbf{i} + y\mathbf{j} = 3\cos\phi\,\mathbf{i} + 3\sin\phi\,\mathbf{j}$$

$$d\mathbf{r} = dx\mathbf{i} + dy\mathbf{j} = (-3\sin\phi\,\mathbf{i} + 3\cos\phi\,\mathbf{j})\, d\phi$$

$$\mathbf{A} = 2y\mathbf{i} - 3x\mathbf{j} + 1.5z\mathbf{k} = 6\sin\phi\,\mathbf{i} - 9\cos\phi\,\mathbf{j} + 1.5z\mathbf{k}$$

$$\begin{aligned}
\mathbf{A} \cdot d\mathbf{r} &= (6\sin\phi\mathbf{i} - 9\cos\phi\mathbf{j} + 1.5z\mathbf{k}) \cdot (-3\sin\phi\mathbf{i} + 3\cos\phi\mathbf{j}) \, d\phi \\
&= \left(-18\sin^2\phi - 27\cos^2\phi\right) d\phi
\end{aligned}$$

$$\begin{aligned}
\int_C \mathbf{A} \cdot d\mathbf{r} &= \int_0^{2\pi} \left(-18\sin^2\phi - 27\cos^2\phi\right) d\phi \\
&= -18\int_0^{2\pi} \sin^2\phi \, d\phi - 27\int_0^{2\pi} \cos^2\phi \, d\phi \\
&= -18\pi - 27\pi \\
&= -45\pi
\end{aligned}$$

Hence, the left and right hand sides are equal and the theorem is verified for this case.

1.50 Make a simple sketch to demonstrate Stokes theorem with sufficient explanations and definitions of the symbols involved.

A sketch should look like Figure 3 with explanation of the symbols seen on the sketch $(S, C, \mathbf{n}, d\sigma, d\mathbf{r})$ as explained in the text and in the answer to exercise 1.48.

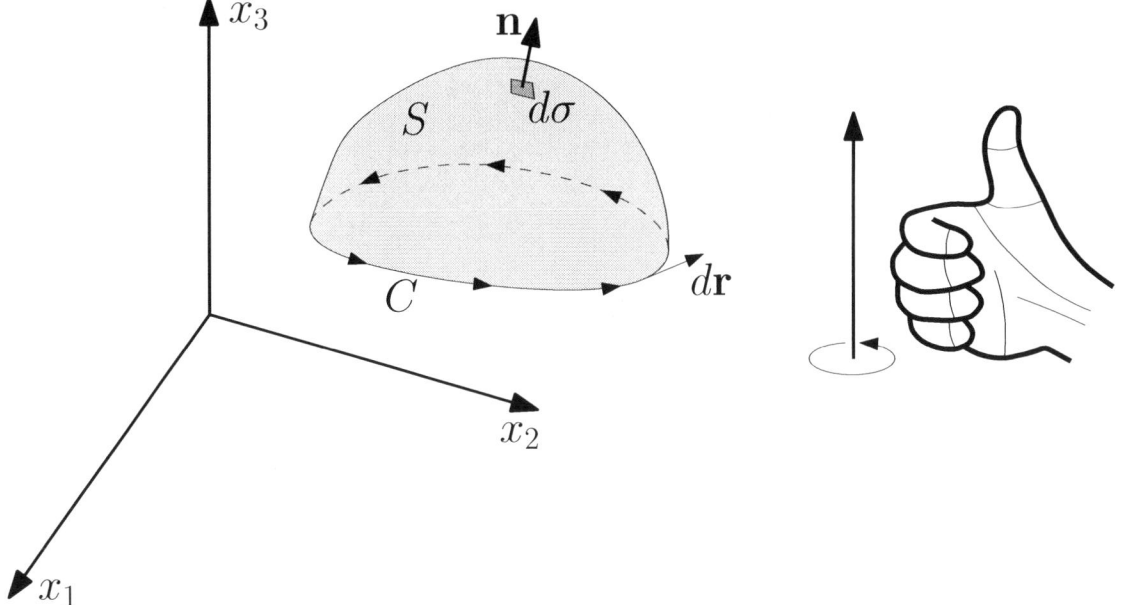

Figure 3: Stokes theorem.

1.51 Give concise definitions for the following terms related to matrices: matrix, square matrix, main diagonal, trailing diagonal, transpose, identity matrix, unit matrix, singular, trace, determinant, cofactor, and inverse.

Matrix: rectangular organized array of mathematical objects such as numbers, variables and functions.

Square matrix: a matrix with the same number of rows and columns.

Main diagonal: the diagonal of a square matrix that runs from top left to bottom right; contains the entries with identical indices, e.g. A_{22}.

Trailing diagonal: the diagonal of a square matrix that runs from top right to bottom left.

Transpose: a matrix obtained by exchanging the rows and columns of the original matrix.

Identity matrix: a square matrix whose all entries are 0 except those on its main diagonal which are 1.

Unit matrix: same as identity matrix.

Singular: a square matrix with no inverse; characterized by having zero determinant.

Trace: sum of the entries on the main diagonal of a square matrix.

Determinant: a scalar associated with a square matrix; may be defined as the sum of the products of each entry of any one of its rows or columns times the cofactor of that entry.

Cofactor: the cofactor of an entry of a square matrix is the determinant obtained from eliminating the row and column of that entry from the parent matrix with a sign given by $(-1)^{i+j}$ where i and j are the indices of the row and column of that entry.

Inverse: the inverse of an $n \times n$ matrix $\mathbf{A}$ is an $n \times n$ matrix $\mathbf{A}^{-1}$ such that their matrix product is the $n \times n$ identity matrix $\mathbf{I}$, i.e. $\mathbf{AA}^{-1} = \mathbf{A}^{-1}\mathbf{A} = \mathbf{I}$.

1.52 Explain the way by which matrices are indexed.

Since matrices are rectangular arrays, they require two indices. The two indices are usually added to the symbol of the matrix as subscripts (e.g. A_{ij}) where the first index labels the rows while the second index labels the columns. The range of the two indices are: $i = 1, \cdots, r$ and $j = 1, \cdots, c$ where i is the row index, r is the number of rows, j is the column index and c is the number of columns.

1.53 How many indices are needed in indexing a 2×3 matrix, an $n \times n$ matrix, and an $m \times k$ matrix? Explain, in each case, why.

In all cases only two indices are needed because matrices are rectangular arrays and hence one index is needed to label the rows and another index is needed to label the columns. The difference between the aforementioned matrices is the range of their indices which identify the number of rows and columns, so for a 2×3 matrix the range of the first index is 2 (i.e. it has 2 rows) and the range of the second index is 3 (i.e. it has 3 columns). Similarly, an $n \times n$ matrix has n rows and n columns while an $m \times k$ matrix has m rows and k columns. So, for an $m \times n$ matrix the range of its first index is m (representing the number of rows) and the range of its second index is n (representing the number of columns).

1.54 Does the order of the matrix indices matter? If so, what is the meaning of changing this order?
The order does matter. By interchanging the indices of a matrix we obtain its transpose.

1.55 Is it possible to write a vector as a matrix? If so, what is the condition that should be imposed on the indices and how many forms a vector can have when it is written as a matrix?
Yes. The condition is that the range of one of its two indices should take only one value, i.e. 1. So, a vector in an nD space can be represented either as a $1 \times n$ matrix (i.e. row matrix) or as an $n \times 1$ matrix (i.e. column matrix). Hence, it can be written in two forms. Each one of these forms is a transpose of the other form.

1.56 Write down the following matrices in a standard rectangular array form using conventional symbols for their entries with a proper indexing: 3×4 matrix $\mathbf{A}$, 1×5 matrix $\mathbf{B}$, 2×2 matrix $\mathbf{C}$, and 3×1 matrix $\mathbf{D}$.

$$\begin{bmatrix} A_{11} & A_{12} & A_{13} & A_{14} \\ A_{21} & A_{22} & A_{23} & A_{24} \\ A_{31} & A_{32} & A_{33} & A_{34} \end{bmatrix}, \begin{bmatrix} B_{11} & B_{12} & B_{13} & B_{14} & B_{15} \end{bmatrix}, \begin{bmatrix} C_{11} & C_{12} \\ C_{21} & C_{22} \end{bmatrix}, \begin{bmatrix} D_{11} \\ D_{21} \\ D_{31} \end{bmatrix}$$

1.57 Give detailed mathematical definitions of the determinant, trace and inverse of matrix, explaining any symbol or technical term involved in these definitions.
The determinant of an $n \times n$ matrix $\mathbf{A}$ is given in Cartesian form by:

$$\det(\mathbf{A}) = \frac{1}{n!} \epsilon_{i_1 \cdots i_n} \epsilon_{j_1 \cdots j_n} A_{i_1 j_1} \ldots A_{i_n j_n}$$

where the indexed ϵ is the permutation symbol and $n!$ symbolizes the factorial of n.
The trace of an $n \times n$ matrix $\mathbf{A}$ is given in Cartesian form by:

$$\text{tr}(\mathbf{A}) = A_{ii} \qquad (i = 1, \ldots, n)$$

1 PRELIMINARIES

where summation over i is assumed.

The inverse of a 3×3 matrix $\mathbf{A}$ is given in Cartesian form by:

$$\left[\mathbf{A}^{-1}\right]_{ij} = \frac{1}{2 \det(\mathbf{A})} \epsilon_{jmn} \epsilon_{ipq} A_{mp} A_{nq}$$

where det stands for determinant and the indexed ϵ is the permutation symbol.

1.58 Find the following matrix multiplications: $\mathbf{AB}$, $\mathbf{BC}$, and $\mathbf{CB}$ where:

$$\mathbf{A} = \begin{bmatrix} 9.6 & 6.3 & -22 \\ -3.8 & 2.5 & 2.9 \\ -6 & 3.2 & 7.5 \end{bmatrix} \quad \mathbf{B} = \begin{bmatrix} -3.8 & -2.0 \\ 4.6 & 11.6 \\ 12.0 & 25.9 \end{bmatrix} \quad \mathbf{C} = \begin{bmatrix} 3 & 8.4 & 61.3 \\ -5 & -33 & 5.9 \end{bmatrix}$$

Answer: By inner product multiplication of the rows of the first matrix by the columns of the second matrix to find the corresponding entries of the product matrix, we obtain:

$$\mathbf{AB} = \begin{bmatrix} 9.6 & 6.3 & -22 \\ -3.8 & 2.5 & 2.9 \\ -6 & 3.2 & 7.5 \end{bmatrix} \begin{bmatrix} -3.8 & -2.0 \\ 4.6 & 11.6 \\ 12.0 & 25.9 \end{bmatrix} = \begin{bmatrix} -271.5 & -515.92 \\ 60.74 & 111.71 \\ 127.52 & 243.37 \end{bmatrix}$$

$$\mathbf{BC} = \begin{bmatrix} -3.8 & -2.0 \\ 4.6 & 11.6 \\ 12.0 & 25.9 \end{bmatrix} \begin{bmatrix} 3 & 8.4 & 61.3 \\ -5 & -33 & 5.9 \end{bmatrix} = \begin{bmatrix} -1.4 & 34.08 & -244.74 \\ -44.2 & -344.16 & 350.42 \\ -93.5 & -753.9 & 888.41 \end{bmatrix}$$

$$\mathbf{CB} = \begin{bmatrix} 3 & 8.4 & 61.3 \\ -5 & -33 & 5.9 \end{bmatrix} \begin{bmatrix} -3.8 & -2.0 \\ 4.6 & 11.6 \\ 12.0 & 25.9 \end{bmatrix} = \begin{bmatrix} 762.84 & 1679.11 \\ -62 & -219.99 \end{bmatrix}$$

1.59 Referring to the matrices $\mathbf{A}$, $\mathbf{B}$ and $\mathbf{C}$ in the previous exercise, find all the permutations (repetitive and non-repetitive) involving two of these three matrices, and classify them into two groups: those which do represent possible matrix multiplication and those which do not.

The number of all permutations is $3 \times 3 = 9$, that is:

$$\begin{array}{ccc} \mathbf{AA} & \mathbf{AB} & \mathbf{AC} \\ \mathbf{BA} & \mathbf{BB} & \mathbf{BC} \\ \mathbf{CA} & \mathbf{CB} & \mathbf{CC} \end{array}$$

1 PRELIMINARIES

Possible matrix multiplication:

$$AA \quad AB$$
$$BC$$
$$CA \quad CB$$

Impossible matrix multiplication: all the other permutations.

1.60 Is matrix multiplication associative? commutative? distributive over matrix addition?

Matrix multiplication is associative and distributive over matrix addition, but it is not commutative.

1.61 Calculate the trace, the determinant, and the inverse (if the inverse exists) of the following matrices:

$$\mathbf{D} = \begin{bmatrix} 3.2 & 2.6 & 1.6 \\ 12.9 & -1.9 & 2.4 \\ -11.9 & 33.2 & -22.5 \end{bmatrix} \qquad \mathbf{E} = \begin{bmatrix} 5.2 & 2.7 & 3.6 \\ -10.4 & -5.4 & -7.2 \\ -31.9 & 13.2 & -23.7 \end{bmatrix}$$

Answer:

$$\operatorname{tr}(\mathbf{D}) = 3.2 - 1.9 - 22.5 = -21.2$$

$$\operatorname{tr}(\mathbf{E}) = 5.2 - 5.4 - 23.7 = -23.9$$

$$\det(\mathbf{D}) = 1211.29$$

$$\det(\mathbf{E}) = 0$$

$$\mathbf{D}^{-1} = \begin{bmatrix} -0.030488 & 0.092150 & 0.007661 \\ 0.216042 & -0.043722 & 0.010699 \\ 0.334907 & -0.113251 & -0.032709 \end{bmatrix} \quad \text{(to 6 decimals)}$$

Since $\det(\mathbf{E}) = 0$, $\mathbf{E}$ has no inverse.

1.62 Which, if any, of the matrices $\mathbf{D}$ and $\mathbf{E}$ in the previous exercise is singular?

$\mathbf{E}$ is singular.

1.63 Select from the Index six terms connected to the special matrices which are defined in § Special Matrices.

Diagonal matrix, identity matrix, singular matrix, transpose of matrix, unit matrix, zero matrix.

Chapter 2
Tensors

2.1 Make a sketch of a rank-2 tensor A_{ij} in a 4D space. What this tensor looks like? The sketch should look like Figure 4.

Figure 4: Rank-2 tensor in 4D space.

It looks like a square matrix (in fact it is representable by a square matrix).

2.2 What are the two main types of notation used for labeling tensors? State two names for each.
(1) Index-free notation, also known as direct or symbolic or Gibbs notation.
(2) Indicial notation, also known as index or component or tensor notation.

2.3 Make a detailed comparison between the two types of notation in the previous question stating any advantages or disadvantages in using one of these notations or the other. In which context each one of these notations is more appropriate to use than the other?
Comparison: the index-free notation (IF) is of geometric nature with no reference

to a particular coordinate system, while the indicial notation (IN) is of algebraic nature with an indication to an underlying coordinate system and basis tensors. Non-indexed bold straight symbols are usually used to represent IF, while indexed light italic symbols are usually used to represent IN. IF is used in general representation while IN is used in specific representation, formulations and calculations.

Advantages and disadvantages: IF is more general and succinct and easier to read than IN, while IN is more specific and informative. IN may be susceptible to some confusion since the same symbol (like A_i) may be used to represent the components as well as the tensor itself. IN may also be more susceptible to error in writing and typesetting due to the presence of indices and may also require more overhead in this regard since writing and typesetting indexed symbols in a legible form usually require extra effort especially when using simple editors, for example, although bold-facing (or using similar notational measures like underlining or using over-arrows) also requires additional effort.

Contexts: IF is recommended for general representation while IN should be used in specific representation that requires the revelation of the underlying structure and indication of the reference frame and basis vectors such as during formulation and calculation. For example, when we talk about a tensor that can be covariant or contravariant or mixed or we want to talk about a tensor whose variance type is irrelevant in that context, then it is more appropriate to use IF notation like **A** because it is general and can represent any variance type, but if we talk specifically about the properties and the rules that apply specifically to one of these variance types (such as covariant type) or we intend to use the tensor symbol in explicit tensor formulation, calculation and development of analytical arguments and proofs then it is more appropriate (and may even be necessary) to use IN notation like A_i for that tensor.

2.4 What is the principle of invariance of tensors and why it is one of the main reasons for the use of tensors in science?

The principle of invariance means that tensor formulations take the same form irrespective of the employed coordinate system and reference frame and hence they are form-invariant. This means that tensors are able to provide objective description of physical phenomena which does not depend on the observer and his coordinate system or reference frame.

2.5 What are the two different meanings of the term "covariant" in tensor calculus?

(1) Invariant (2) as opposite to contravariant.

2.6 State the type of each one of the following tensors considering the number and position of indices (i.e. covariant, contravariant, rank, scalar, vector, etc.):

$$a^i \qquad B_i^{jk} \qquad f \qquad b_k \qquad C^{ji}$$

Answer:
a^i is contravariant of type $(1,0)$, rank-1, vector.
B_i^{jk} is mixed of type $(2,1)$, rank-3, tensor.
f is scalar of type $(0,0)$, rank-0.
b_k is covariant of type $(0,1)$, rank-1, vector.
C^{ji} is contravariant of type $(2,0)$, rank-2, tensor.

2.7 Define the following technical terms which are related to tensors: term, expression, equality, order, rank, zero tensor, unit tensor, free index, dummy index, covariant, contravariant, and mixed.
Tensor term: product of tensors including scalars and vectors.
Tensor expression: algebraic sum of tensor terms.
Tensor equality: equality of two tensor terms and/or expressions.
Tensor order: indicator to the total number of tensor indices.
Tensor rank: indicator to the number of free indices of tensor.
Zero tensor: a tensor whose all components are zero.
Unit tensor: a tensor whose all components are zero except those with identical values of all indices which are assigned the value 1.
Free index: index that occurs only once in a tensor term.
Dummy index: index that occurs twice in a tensor term.
Covariant tensor: tensor whose all indices are subscripts.
Contravariant tensor: tensor whose all indices are superscripts.
Mixed tensor: tensor with both subscripts and superscript indices.

2.8 Which of the following is a scalar, vector or rank-2 tensor: temperature, stress, cross product of two vectors, dot product of two vectors, and rate of strain?
Scalar, rank-2 tensor, vector, scalar, rank-2 tensor.

2.9 What is the number of entries of a rank-0 tensor in a 2D space and in a 5D space? What is the number of entries of a rank-1 tensor in these spaces?

The number of entries of a rank-0 tensor is 1 in any space of any dimension. The number of entries of a rank-1 tensor in these spaces is 2 and 5 respectively.

2.10 What is the difference between the order and rank of a tensor considering the different conventions in this regard?

The order represents the total number of indices including dummy indices, while the rank represents the number of free indices only. It is common to use "order" to mean what we call "rank".

2.11 What is the number of entries of a rank-3 tensor in a 4D space? What is the number of entries of a rank-4 tensor in a 3D space?

The number of components of a tensor is given by the following formula:

$$N = D^R$$

where N is the number of components, D is the space dimension and R is the tensor rank. Hence, we have for the two cases in the question:

$4^3 = 64$.

$3^4 = 81$.

2.12 Describe direct and inverse coordinate transformations between spaces and write the generic equations for these transformations.

Direct transformation is a transformation from original to subsidiary coordinates while inverse transformation is a transformation from subsidiary to original coordinates. Mathematically, direct and inverse transformations can be represented respectively by the following equations:

$$\bar{x}^i = \bar{x}^i(x^1, x^2, \ldots, x^n)$$
$$x^i = x^i(\bar{x}^1, \bar{x}^2, \ldots, \bar{x}^n)$$

where $i = 1, 2, \ldots, n$ with n being the space dimension and the bars mark subsidiary coordinates.

2.13 What are proper and improper transformations? Draw a simple sketch to demonstrate them.

Proper transformation preserves system handedness (right or left) while improper transformation reverses system handedness. The sketch should look like Figure 5.

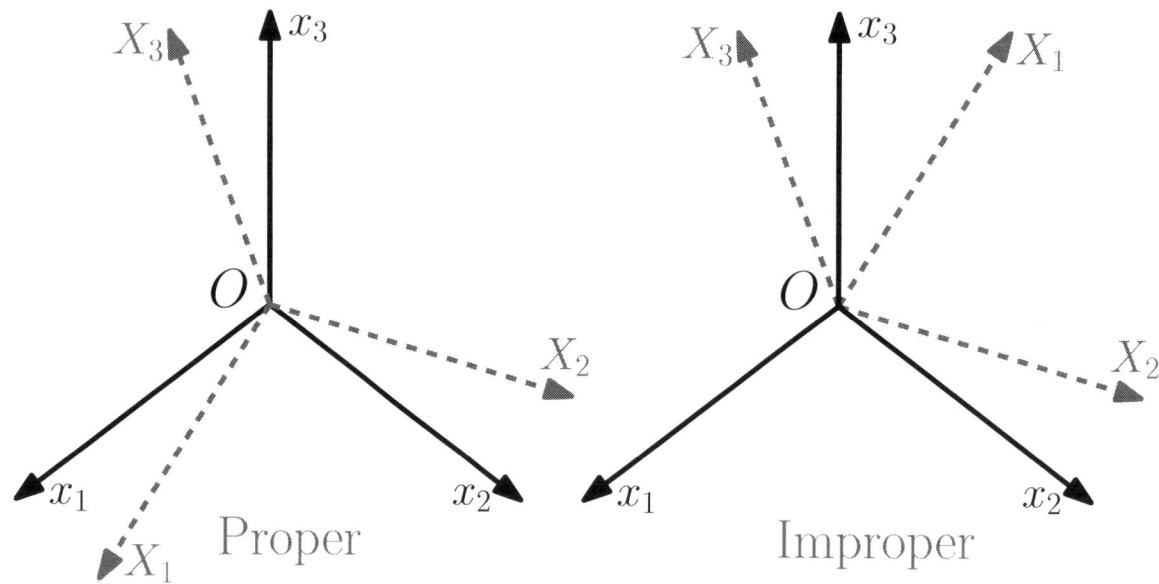

Figure 5: Proper and improper transformations.

2.14 Define the following terms related to permutation of indices: permutation, even, odd, parity, and transposition.

Permutation: arrangement of indices in a certain order.

Even: a permutation resulting from an even number of single-step exchanges (transpositions) of neighboring indices starting from a presumed original arrangement of these indices.

Odd: a permutation resulting from an odd number of transpositions.

Parity: being odd or even.

Transposition: single-step exchange of neighboring indices.

2.15 Find all the permutations of the following four letters assuming no repetition: (i, j, k, l).
Answer:
$$\begin{array}{cccccc} ijkl & ijlk & ikjl & iklj & iljk & ilkj \\ jikl & jilk & jkil & jkli & jlik & jlki \\ kijl & kilj & kjil & kjli & klij & klji \\ lijk & likj & ljik & ljki & lkij & lkji \end{array}$$

2.16 Give three even permutations and three odd permutations of the symbols $(\alpha, \beta, \gamma, \delta)$ in the stated order.
Even permutations: $\alpha\beta\gamma\delta$, $\beta\gamma\alpha\delta$, $\alpha\gamma\delta\beta$.
Odd permutations: $\beta\alpha\gamma\delta$, $\alpha\gamma\beta\delta$, $\alpha\beta\delta\gamma$.

2 TENSORS

2.17 Discuss all the similarities and differences between free and dummy indices.

Answer:
(a) Free indices (FI) and dummy indices (DI) are both ordinary tensor indices and hence they are subject to the same general rules of tensor indices (e.g. being either covariant or contravariant).
(b) Both FI and DI range over certain number of dimensions representing space dimensionality.
(c) FI occur once in any tensor term and DI occur twice.
(d) DI imply summation (when this convention is adopted) but FI do not.
(e) DI are restricted to their terms and hence they can occur only in some terms of tensor expressions and equalities while FI should have extended presence in all terms.
(f) DI can be replaced in individual terms (as long as the new label is not used in that term) but FI cannot although FI can be replaced in all terms if the new label is not in use in that context.
(g) When DI are present in more than one term of tensor expressions and equalities they can be named differently in each term but FI should be named uniformly in all terms.
(h) FI count in tensor rank and order but DI count only in tensor order.
(i) DI can be present in scalar quantity (when all indices are contracted) but FI can not.

2.18 What is the maximum number of repetitive indices that can occur in each term of a legitimate tensor expression?
Two, although the violation of this restriction is common in the literature. Yes, this condition strictly applies to dummy indices where the stated rules of dummy indices apply only to twice-repeated indices.

2.19 How many components are represented by each one of the following assuming a 4D space?

$$A_i^{jk} \qquad f+g \qquad C^{mn} - D^{mn} \qquad 5D_k + 4A_k = B_k$$

Answer:
$$4^3 = 64 \qquad 4^0 = 1 \qquad 4^2 = 16 \qquad 4^1 = 4$$

2.20 What is the "summation convention"? To what type of indices this convention applies?

The summation convention states that a twice-repeated index in a tensor term implies summation over the index range. This convention applies only to dummy (also known as bound) indices.

2.21 Is it always the case that the summation convention applies when an index is repeated? If not, what precaution should be taken to avoid ambiguity and confusion?
Not always. Sometimes a twice-repeated index refers to the components with identical value of indices, e.g. A_i^i to mean the tensor components with the same value of upper and lower indices like A_1^1 or A_3^3. There are several types of precaution that can be taken when this happens; these precautions include parenthesizing the repetitive indices which are not subject to the summation convention (e.g. $A_{(i)}^{(i)}$), or using upper case letters for labeling these indices (e.g. A_I^I), or adding comments like "no sum" to indicate that the summation convention does not apply in these cases.

2.22 In which cases a pair of dummy indices should be of different variance type (i.e. one upper and one lower)? In what type of coordinate systems these repeated indices can be of the same variance type and why?
This applies in general. The exception is orthonormal Cartesian systems where the variance type of indices is irrelevant and hence they can be both upper or both lower since the basis vector set for these types is the same.

2.23 What are the rules that the free indices should obey when they occur in the terms of tensor expressions and equalities?
(a) Each term should have the same number of free indices.
(b) A free index should have the same variance type in all terms.
(c) Each term should have the same set of free indices, e.g. all terms should have i, j, k and hence it is not allowed to have one term with i, j, k set and another term with i, j, n set.
(d) The free indices should have the same arrangement.
(e) Each index should have the same range (i.e. space dimensionality) in all terms, and hence the index i in $A_i + B_i$ expression should have identical range in both terms. This condition may also be imposed on the indices as a whole and hence in $A_i^{jk} + B_i^{jk}$ expression all the three indices i, j, k should have the same range $1, \cdots, n$.

2.24 What is illegitimate about the following tensor expressions and equalities considering

2 TENSORS

in your answer all the possible violations?

$$A_{ij} + B_{ij}^k \qquad C^n - D_n = B_m \qquad A_j^i = A_i^j \qquad A_j = f$$

(1) Different number of free indices (rule a).
(2) Different variance type and different set of free indices (rules b and c).
(3) Different variance type (rule b).
(4) Different number of free indices (rule a).

2.25 Which of the following tensor expressions and equalities is legitimate and which is illegitimate?

$$B^i + C_j^{ij} \qquad A_i - B_i^k \qquad C^m + D^m = B_{mm}^m \qquad B_k^i = A_k^i$$

State in each illegitimate case all the reasons for illegitimacy.
Legitimate.
Illegitimate: violation of rule a (i.e. different number of free indices).
Illegitimate: violation of the rules of labeling indices since B_{mm}^m has three repetitive indices with no obvious sensible meaning.
Legitimate.

2.26 Which is right and which is wrong of the following tensor equalities?

$$\partial_n A_n = A_n \partial_n \qquad [\mathbf{B}]_k + [\mathbf{D}]_k = [\mathbf{B} + \mathbf{D}]_k \qquad ab = ba \qquad A^{ij} M_{kl} = M_{kl} A^{ji}$$

Explain in each case why the equality is right or wrong.
Wrong: operators do not commute with their operands.
Right: indexing is distributive.
Right: ordinary multiplication of scalars is commutative.
Wrong: tensor multiplication is not commutative; however, this can be right if it represents the equality of components.

2.27 Choose from the Index six entries related to the general rules that apply to the mathematical expressions and equalities in tensor calculus.
Some of possible entries are: distributive, bound index, covariant index, contravariant index, dummy index, free index, indexed square bracket, indicial notation, indicial structure, rank of tensor, summation convention, tensor term, variance type.

2.28 Give at least two examples of tensors used in mathematics, science and engineering for each one of the following ranks: 0, 1, 2 and 3.
Rank-0: mass, temperature.
Rank-1: acceleration, force.
Rank-2: stress, rate of strain.
Rank-3: Levi-Civita tensor in 3D space, piezoelectric moduli tensor.

2.29 State the special names given to the rank-0 and rank-1 tensors.
Scalars and vectors respectively.

2.30 What is the difference, if any, between rank-2 tensors and matrices?
Rank-2 tensors are abstract mathematical objects that can be represented by matrices, so matrices is a form for representing and symbolizing these abstract objects.

2.31 Is the following statement correct? If not, re-write it correctly: "all rank-0 tensors are vectors and vice versa, and all rank-1 tensors are scalars and vice versa".
Wrong. Corrected: "Rank-0 tensors are scalars and rank-1 tensors are vectors, but not all scalars (in generic sense) are rank-0 tensors and not all vectors (in generic sense) are rank-1 tensors".

2.32 Give clear and detailed definitions of scalars and vectors and compare them. What is common and what is different between the two?
"Scalars" may be used generically to mean quantities that have a single component (i.e. they are fully identified by their magnitude and algebraic sign) irrespective of satisfying the principle of invariance or not. Similarly, "vectors" may be used generically to mean quantities that have exactly n components in nD space (i.e. they are fully identified by their magnitude and a single direction in space) irrespective of satisfying the principle of invariance or not. However, it is common in the technical terminology of tensor calculus to use "scalars" to mean rank-0 tensors and "vectors" to mean rank-1 tensors where "tensors" mean objects that satisfy the principle of invariance and hence they should satisfy this principle. Accordingly, when this terminology is employed we will have "scalars" and "pseudo scalars" and "vectors" and "pseudo vectors", i.e. pseudo tensors of rank-0 and pseudo tensors of rank-1 respectively. Therefore, one should be careful when reading the literature of tensor calculus to distinguish between the generic and the more technical sense of these terms to avoid confusion and misunderstanding.
Common: both are mathematical objects, both satisfy the principle of invariance

when used technically.

Different: in nD space scalars have $n^0 = 1$ component while vectors have $n^1 = n$ components.

2.33 Make a simple sketch of the nine unit dyads associated with the double directions of rank-2 tensors in a 3D space.

A sketch should look like Figure 6.

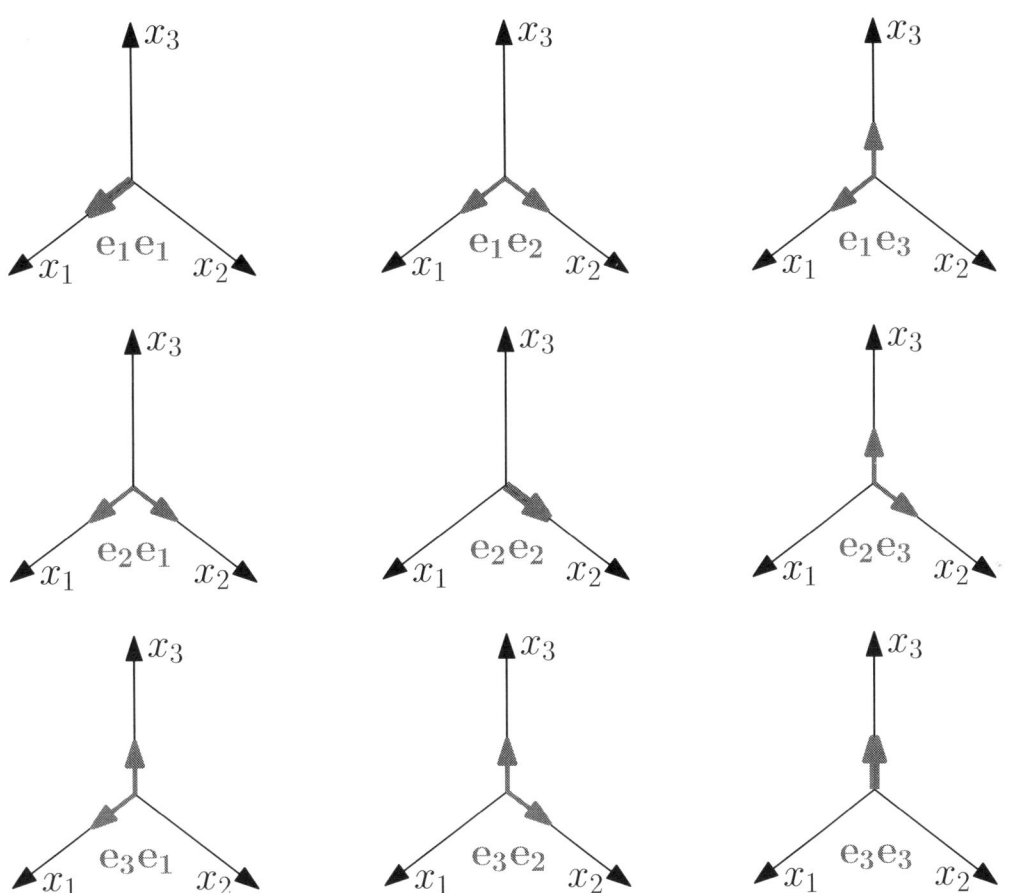

Figure 6: Nine unit dyads associated with the double directions of rank-2 tensors in a 3D space.

2.34 Name three of the scientific disciplines that heavily rely on tensor calculus notation and techniques.

Differential geometry, fluid dynamics, rheology.

2.35 What are the main features of tensor calculus that make it very useful and successful in mathematical, scientific and engineering applications.

Ability to model multi-dimensional problems in compact and efficient way, ability to do complex mathematical manipulations and formulations in relatively easy way, ability to provide form-invariant scientific formulations.

2.36 Why tensor calculus is used in the formulation and presentation of the laws of physics?
Because tensor formulations are form-invariant and hence they provide universally valid description and modeling of physical phenomena irrespective of coordinate systems, observers and their frames of reference.

2.37 Give concise definitions for the covariant and contravariant types of tensor.
Covariant tensors are labeled with lower indices, their components refer to contravariant basis tensor set, e.g. $A_{i_1 \cdots i_n}$ refers to the contravariant basis tensor $\mathbf{E}^{i_1} \cdots \mathbf{E}^{i_n}$.
Contravariant tensors are labeled with upper indices, their components refer to covariant basis tensor set, e.g. $A^{i_1 \cdots i_n}$ refers to the covariant basis tensor $\mathbf{E}_{i_1} \cdots \mathbf{E}_{i_n}$.

2.38 Describe how the covariant and contravariant types are notated and how they differ in their transformation between coordinate systems.
Covariant tensors are notated with lower indices (e.g. $A_{i_1 \cdots i_n}$), they transform according to the following rule:

$$\bar{A}_{i_1 \cdots i_n} = \frac{\partial x^{j_1}}{\partial \bar{x}^{i_1}} \cdots \frac{\partial x^{j_n}}{\partial \bar{x}^{i_n}} A_{j_1 \cdots j_n}$$

Contravariant tensors are notated with upper indices (e.g. $A^{i_1 \cdots i_n}$), they transform according to the following rule:

$$\bar{A}^{i_1 \cdots i_n} = \frac{\partial \bar{x}^{i_1}}{\partial x^{j_1}} \cdots \frac{\partial \bar{x}^{i_n}}{\partial x^{j_n}} A^{j_1 \cdots j_n}$$

2.39 Give examples of tensors used in mathematics and science which are covariant and other examples which are contravariant.
In fact, any tensor can be put into covariant or contravariant form depending on the adopted variance type of the basis vector set of the given coordinate system; so if we use the contravariant/covariant basis set of the system we write our tensors in covariant/contravariant form. Hence, virtually all the tensors that have been mentioned in the text can be covariant or contravariant. Examples: permutation tensor (ϵ_{ijk}, ϵ^{ijk}), Kronecker delta tensor (δ_{ij}, δ^{ij}), stress tensor (S_{ij}, S^{ij}), and so on.

2.40 Write the mathematical transformation rules of the following tensors: A_{ijk} to $\bar{A}_{rst}$

and B^{mn} to $\bar{B}^{pq}$.

Answer:
$$\bar{A}_{rst} = \frac{\partial x^i}{\partial \bar{x}^r}\frac{\partial x^j}{\partial \bar{x}^s}\frac{\partial x^k}{\partial \bar{x}^t} A_{ijk}$$

$$\bar{B}^{pq} = \frac{\partial \bar{x}^p}{\partial x^m}\frac{\partial \bar{x}^q}{\partial x^n} B^{mn}$$

2.41 Explain how mixed type tensors are defined and notated in tensor calculus.

Mixed tensors have covariant indices and contravariant indices at the same time and hence they are tensors with upper and lower indices, e.g. B^{ij}_k. Their components refer to tensor basis set of mixed type opposite in variance type to their indices, e.g. $A^i{}_j$ refers to the basis $\mathbf{E}_i \mathbf{E}^j$. Accordingly, they are transformed covariantly with respect to their covariant indices and contravariantly with respect to their contravariant indices.

2.42 Write the mathematical rule for transforming the mixed type tensor $D^{ij}{}_{klm}$ to $\bar{D}^{pq}{}_{rst}$.

Answer:
$$\bar{D}^{pq}{}_{rst} = \frac{\partial \bar{x}^p}{\partial x^i}\frac{\partial \bar{x}^q}{\partial x^j}\frac{\partial x^k}{\partial \bar{x}^r}\frac{\partial x^l}{\partial \bar{x}^s}\frac{\partial x^m}{\partial \bar{x}^t} D^{ij}{}_{klm}$$

2.43 From the Index, find all the terms that start with the word "Mixed" and are related specifically to tensors of rank 2.

Mixed Kronecker δ, mixed metric tensor.

2.44 Express the following tensors in indicial notation: a rank-3 covariant tensor **A**, a rank-4 contravariant tensor **B**, a rank-5 mixed type tensor **C** which is covariant in ij indices and contravariant in kmn indices where the indices are ordered as $ikmnj$.

Answer:
$$A_{ijk} \qquad B^{ijkl} \qquad C_i{}^{kmn}{}_j$$

2.45 Write step-by-step the mathematical transformations of the following tensors: A_{ij} to $\bar{A}_{rs}$, B^{lmn} to $\bar{B}^{pqr}$, $C^{ij}{}_{mn}$ to $\bar{C}^{pq}{}_{rs}$ and $D_m{}^{kl}$ to $\bar{D}_r{}^{st}$.

Answer:
$$\bar{A} = A$$
$$\bar{A}_{rs} = A_{ij}$$
$$\bar{A}_{rs} = \frac{\partial}{\partial}\frac{\partial}{\partial} A_{ij}$$
$$\bar{A}_{rs} = \frac{\partial}{\partial x^r}\frac{\partial}{\partial x^s} A_{ij}$$

$$\bar{A}_{rs} = \frac{\partial}{\partial \bar{x}^r} \frac{\partial}{\partial \bar{x}^s} A_{ij}$$

$$\bar{A}_{rs} = \frac{\partial x^i}{\partial \bar{x}^r} \frac{\partial x^j}{\partial \bar{x}^s} A_{ij}$$

$$\bar{B} = B$$

$$\bar{B}^{pqr} = B^{lmn}$$

$$\bar{B}^{pqr} = \frac{\partial}{\partial} \frac{\partial}{\partial} \frac{\partial}{\partial} B^{lmn}$$

$$\bar{B}^{pqr} = \frac{\partial x^p}{\partial} \frac{\partial x^q}{\partial} \frac{\partial x^r}{\partial} B^{lmn}$$

$$\bar{B}^{pqr} = \frac{\partial \bar{x}^p}{\partial} \frac{\partial \bar{x}^q}{\partial} \frac{\partial \bar{x}^r}{\partial} B^{lmn}$$

$$\bar{B}^{pqr} = \frac{\partial \bar{x}^p}{\partial x^l} \frac{\partial \bar{x}^q}{\partial x^m} \frac{\partial \bar{x}^r}{\partial x^n} B^{lmn}$$

$$\bar{C} = C$$

$$\bar{C}^{pq}_{\ rs} = C^{ij}_{\ mn}$$

$$\bar{C}^{pq}_{\ rs} = \frac{\partial}{\partial} \frac{\partial}{\partial} \frac{\partial}{\partial} \frac{\partial}{\partial} C^{ij}_{\ mn}$$

$$\bar{C}^{pq}_{\ rs} = \frac{\partial x^p}{\partial} \frac{\partial x^q}{\partial} \frac{\partial}{\partial x^r} \frac{\partial}{\partial x^s} C^{ij}_{\ mn}$$

$$\bar{C}^{pq}_{\ rs} = \frac{\partial \bar{x}^p}{\partial} \frac{\partial \bar{x}^q}{\partial} \frac{\partial}{\partial \bar{x}^r} \frac{\partial}{\partial \bar{x}^s} C^{ij}_{\ mn}$$

$$\bar{C}^{pq}_{\ rs} = \frac{\partial \bar{x}^p}{\partial x^i} \frac{\partial \bar{x}^q}{\partial x^j} \frac{\partial x^m}{\partial \bar{x}^r} \frac{\partial x^n}{\partial \bar{x}^s} C^{ij}_{\ mn}$$

$$\bar{D} = D$$

$$\bar{D}_r^{\ st} = D_m^{\ kl}$$

$$\bar{D}_r^{\ st} = \frac{\partial}{\partial} \frac{\partial}{\partial} \frac{\partial}{\partial} D_m^{\ kl}$$

$$\bar{D}_r^{\ st} = \frac{\partial}{\partial x^r} \frac{\partial x^s}{\partial} \frac{\partial x^t}{\partial} D_m^{\ kl}$$

$$\bar{D}_r^{\ st} = \frac{\partial}{\partial \bar{x}^r} \frac{\partial \bar{x}^s}{\partial} \frac{\partial \bar{x}^t}{\partial} D_m^{\ kl}$$

$$\bar{D}_r^{\ st} = \frac{\partial x^m}{\partial \bar{x}^r} \frac{\partial \bar{x}^s}{\partial x^k} \frac{\partial \bar{x}^t}{\partial x^l} D_m^{\ kl}$$

2.46 What is the relation between the rank and the (m,n) type of a tensor?

2 TENSORS

Rank $= m + n$.

2.47 Write, in indicial notation, the following tensors: **A** of type $(0,4)$, **B** of type $(3,1)$, **C** of type $(0,0)$, **D** of type $(3,4)$, **E** of type $(2,0)$ and **F** of type $(1,1)$.
Answer:
$$A_{ijkl} \qquad B^{ijk}_l \qquad C \qquad D^{ijk}_{qrst} \qquad E^{ij} \qquad F^i_j$$

2.48 What is the rank of each one of the tensors in the previous question? Are there tensors among them which may not have been notated properly?
Rank: 4, 4, 0, 7, 2, 2.
Yes, C which should not be boldfaced in the question.

2.49 Which tensor provides the link between the covariant and contravariant types of a given tensor **D**?
Metric tensor.

2.50 What coordinate system(s) in which the covariant and contravariant types of a tensor do not differ? What is the usual tensor notation used in this case?
Orthonormal Cartesian.
Subscripts are usually used to index the tensors.

2.51 Define in detail, qualitatively and mathematically, the covariant and contravariant types of the basis vectors of a general coordinate system explaining all the symbols used in your definition.
Covariant basis vectors are the tangents to the coordinate curves which are the curves along which only one coordinate varies, that is:
$$\mathbf{E}_i = \frac{\partial \mathbf{r}}{\partial x^i}$$
where $\mathbf{E}_i$ is the covariant basis vector, $\mathbf{r}$ is the position vector in Cartesian coordinates and x^i is a general coordinate.
Contravariant basis vectors are the gradients to the coordinate surfaces which are the surfaces over which only one coordinate is constant, that is:
$$\mathbf{E}^i = \nabla x^i$$
where $\mathbf{E}^i$ is the contravariant basis vector and ∇ is the nabla operator of a Cartesian system as defined previously.

2.52 Is it necessary that the basis vectors of the previous exercise are mutually orthogonal and/or of unit length?

No. The basis vectors in general coordinate systems can vary in their relative orientation and in magnitude and hence they may not be mutually orthogonal and/or of unit length.

2.53 Is the following statement correct? "A superscript in the denominator of partial derivatives is equivalent to a superscript in the numerator". Explain why.

No. A superscript in the denominator of partial derivatives is equivalent to a **subscript** in the numerator.

2.54 What is the reciprocity relation that links the covariant and contravariant basis vectors? Express this relation mathematically.

The reciprocity relation is given by:

$$\mathbf{E}_i \cdot \mathbf{E}^j = \delta_i^j$$

It represents the fact that the covariant and contravariant basis vectors are reciprocal systems, i.e. their dot product is zero (and hence they are perpendicular) when they have different indices and their dot product is 1 when they have identical indices.

2.55 What is the interpretation of the reciprocity relation?

The interpretation is that the dot product of two basis vectors of opposite variance type with different indices is zero and hence they are perpendicular, while the dot product of two basis vectors of opposite variance type with identical indices is unity.

2.56 Are the covariant and contravariant forms of a specific tensor **A** represent the same mathematical object? If so, in what sense they are equal from the perspective of different coordinate system bases?

Yes. In the sense that both forms represent the same object in space although the components are different because the two forms refer to different basis vector sets of the same coordinate system, i.e. the covariant form refers to the contravariant basis set of the system while the contravariant form refers to the covariant basis set of the system.

2.57 Correct, if necessary, the following statement: "A tensor of any rank (≥ 1) can be represented covariantly using contravariant basis tensors of that rank, or contravariantly

2 TENSORS

using contravariant basis tensors, or in a mixed form using a mixed basis of the same type".

This statement is wrong. Corrected: A tensor of any rank (≥ 1) can be represented covariantly using contravariant basis tensors of that rank, or contravariantly using covariant basis tensors, or in a mixed form using a mixed basis of the opposite type (for mixed form the rank should be ≥ 2).

2.58 Make corrections, if needed, to the following equations assuming a general curvilinear coordinate system where, in each case, all the possible ways of correction should be considered:

$$\mathbf{B} = B^i \mathbf{E}^i \quad \mathbf{M} = M_{ij} \mathbf{E}^i \quad \mathbf{D} = D^i \mathbf{E}^i \mathbf{E}_j \quad \mathbf{C} = C^i \mathbf{E}_j \quad \mathbf{F} = F^n \mathbf{E}_n \quad \mathbf{T} = T^{rs} \mathbf{E}_s \mathbf{E}_r$$

Answer:
$\mathbf{B} = B_i \mathbf{E}^i$ or $\mathbf{B} = B^i \mathbf{E}_i$.
$\mathbf{M} = M_{ij} \mathbf{E}^i \mathbf{E}^j$ or $\mathbf{M} = M_i \mathbf{E}^i$.
$\mathbf{D} = D_i^j \mathbf{E}^i \mathbf{E}_j$ or $\mathbf{D} = D^{ij} \mathbf{E}_i \mathbf{E}_j$ or $\mathbf{D} = D^i \mathbf{E}_i$ or $\mathbf{D} = D_i \mathbf{E}^i$.
$\mathbf{C} = C^i \mathbf{E}_i$ or $\mathbf{C} = C^j \mathbf{E}_j$.
$\mathbf{F} = F^n \mathbf{E}_n$ is correct.
$\mathbf{T} = T^{rs} \mathbf{E}_r \mathbf{E}_s$ or $\mathbf{T} = T^{sr} \mathbf{E}_s \mathbf{E}_r$.

2.59 What is the technical term used to label the following objects: $\mathbf{E}^i \mathbf{E}^j$, $\mathbf{E}_i \mathbf{E}_j$, $\mathbf{E}_i \mathbf{E}^j$ and $\mathbf{E}^i \mathbf{E}_j$? What they mean?

Dyads. They represent basis tensors of rank-2; each represents double direction in space based on the underlying basis vector sets.

2.60 What sort of tensor components that the objects in the previous question should be associated with?

Covariant, contravariant, mixed, mixed. All should be of rank-2.

2.61 What is the difference between true and pseudo vectors? Which of these is called axial and which is called polar?

A true vector keeps its direction in space following a reflection of the coordinate system that changes its handedness, while a pseudo vector reverses its direction following this operation. True vectors are called polar while pseudo vectors are called axial.

2.62 Make a sketch demonstrating the behavior of true and pseudo vectors.

A sketch should look like Figure 7.

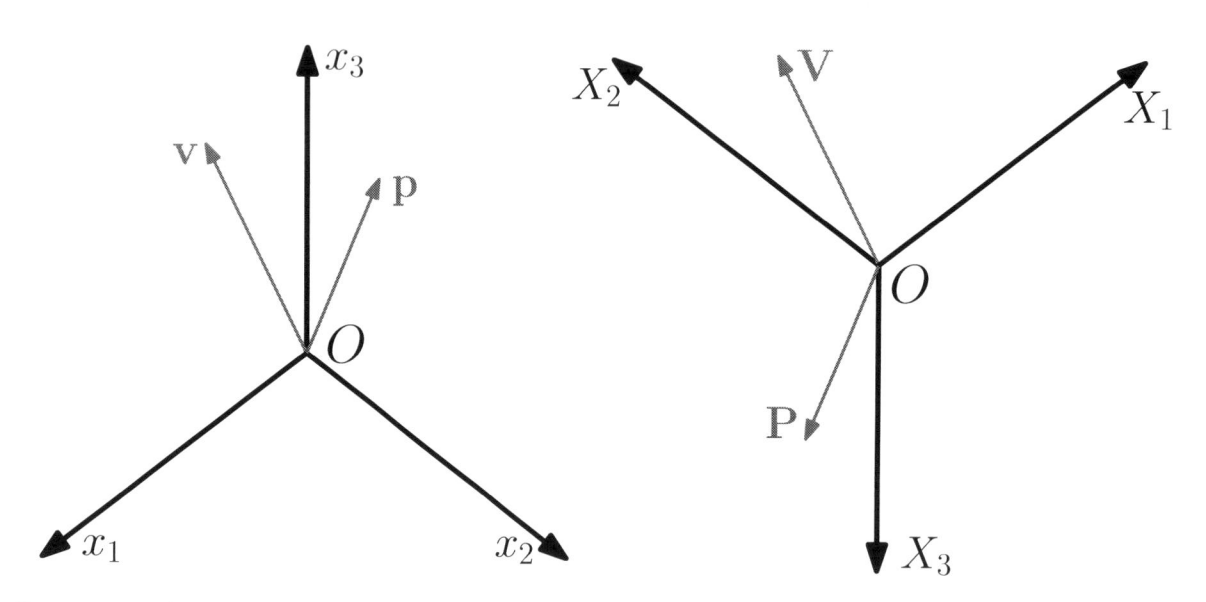

Figure 7: The behavior of a true vector (**v** and **V**) and a pseudo vector (**p** and **P**) following a reflection of the coordinate system.

2.63 Is the following statement correct? "The terms of tensor expressions and equations should be uniform in their true and pseudo type". Explain why.

Yes, it is correct. Because true and pseudo tensors are different in nature and hence they cannot be added algebraically. For example, if we add a true vector to a pseudo vector and the coordinate system is reflected with change of its handedness, part of the resultant vector will keep its direction while the other part will reverse its direction and hence the resultant vector will not represent its original state in a sensible way because it neither kept its direction like a true vector nor reversed its direction like a pseudo vector.

2.64 There are four possibilities for the direct product of two tensors of true and pseudo types. Discuss all these possibilities with respect to the type of the tensor produced by this operation and if it is true or pseudo. Also discuss in detail the cross product and curl operations from this perspective.

We have four possibilities:

$$T \times T = T \qquad P \times P = T \qquad T \times P = P \qquad P \times T = P$$

where T and P stand for true and pseudo respectively and the sign $\times$ represents direct product.

Since cross product and curl operations enclose, by definition, a permutation tensor which is a pseudo tensor, each one of these operations adds an extra pseudo factor to the product.

2.65 Give examples for the true and pseudo types of scalars, vectors and rank-2 tensors.
True scalar: dot product of two true vectors.
True vector: acceleration.
True rank-2 tensor: Kronecker delta tensor.
Pseudo scalar: dot product of a true vector with a pseudo vector.
Pseudo vector: cross product of two true vectors.
Pseudo rank-2 tensor: rank-2 permutation tensor ϵ_{ij}.

2.66 Explain, in words and equations, the meaning of absolute and relative tensors. Do these intersect in some cases with true and pseudo tensors (at least according to some conventions)?
Absolute tensor has no Jacobian factor in its transformation equation, while relative tensor has such a factor. Mathematically, the transformation equation for these two types of tensor is:

$$\bar{A}^{ij...k}{}_{lm...n} = \left|\frac{\partial x}{\partial \bar{x}}\right|^w \frac{\partial \bar{x}^i}{\partial x^a}\frac{\partial \bar{x}^j}{\partial x^b}\cdots\frac{\partial \bar{x}^k}{\partial x^c}\frac{\partial x^d}{\partial \bar{x}^l}\frac{\partial x^e}{\partial \bar{x}^m}\cdots\frac{\partial x^f}{\partial \bar{x}^n}A^{ab...c}{}_{de...f}$$

where $w = 0$ for absolute tensor and $w \neq 0$ for relative tensor.
Yes, they intersect since pseudo tensor is a relative tensor with $w = -1$.

2.67 What "Jacobian" and "weight" mean in the context of absolute and relative tensors?
The "Jacobian" J of a transformation is the determinant of the Jacobian matrix $\mathbf{J}$ which is an $n \times n$ matrix formed of all the possible n^2 partial derivatives of the original coordinates with respect to the transformed coordinates, where n is the space dimension, that is:

$$J = \det(\mathbf{J}) = \begin{vmatrix} \frac{\partial x^1}{\partial \bar{x}^1} & \frac{\partial x^1}{\partial \bar{x}^2} & \cdots & \frac{\partial x^1}{\partial \bar{x}^n} \\ \frac{\partial x^2}{\partial \bar{x}^1} & \frac{\partial x^2}{\partial \bar{x}^2} & \cdots & \frac{\partial x^2}{\partial \bar{x}^n} \\ \vdots & \vdots & \ddots & \vdots \\ \frac{\partial x^n}{\partial \bar{x}^1} & \frac{\partial x^n}{\partial \bar{x}^2} & \cdots & \frac{\partial x^n}{\partial \bar{x}^n} \end{vmatrix}$$

The "weight" in this context is the power index w of the Jacobian in the transformation equation as given in the answer of the previous exercise.

2.68 Someone stated: "$\mathbf{A}$ is a tensor of type $(2, 4, -1)$". What these three numbers refer to?

This means that **A** has two contravariant free indices, 4 covariant free indices and its weight is $w = -1$. Hence, **A** is a rank-6 mixed pseudo tensor.

2.69 What is the type of the tensor in the previous exercise from the perspectives of lower and upper indices and absolute and relative tensors? What is the rank of this tensor?
Mixed type, relative tensor of weight $w = -1$ and hence it is pseudo tensor.
The rank is: $2 + 4 = 6$.

2.70 What is the weight of a tensor **A** produced from multiplying a tensor of weight -1 by a tensor of weight 2? Is **A** relative or absolute? Is it true or not?
Weight: $w = -1 + 2 = 1$.
Since $w \neq 0$, it is relative.
Since $w \neq -1$ it is not pseudo and hence it is true in this sense although it may be called tensor density because $w = 1$ (the conventions and terminologies differ in this regard).

2.71 Define isotropic and anisotropic tensors and give examples for each using tensors of different ranks.
The components of isotropic tensors do not change under proper rotational transformations, while the components of anisotropic tensors do change. Examples of isotropic tensor are Kronecker delta tensor and permutation tensor. Tensors that represent physical properties, such as stress or velocity gradient, are usually anisotropic.

2.72 What is the state of the inner and outer products of two isotropic tensors?
They are isotropic.

2.73 Why if a tensor equation is valid in a particular coordinate system it should also be valid in all other coordinate systems under admissible coordinate transformations? Use the isotropy of the zero tensor in your explanation.
The zero tensor is invariant under all transformations (proper and improper), hence if an equation is valid in a given coordinate system then it should be valid in all coordinate systems since the zero tensor that this equation represents when all its terms are moved to one side (e.g. when we write $\mathbf{A} = \mathbf{B}$ as $\mathbf{A} - \mathbf{B} = \mathbf{0}$) is invariant. In fact, this argument is not limited to equations of isotropic tensors since the zero tensor is invariant under improper as well as proper transformations because changing the system handedness by reflection will not change the value of any component of the zero tensor since $+0 = -0$.

2 TENSORS   43

2.74 Define "symmetric" and "anti-symmetric" tensors and write the mathematical condition that applies to each assuming a rank-2 tensor.

The components of symmetric tensors are invariant under exchange of indices, while the components of anti-symmetric tensors reverse their sign under this exchange. For rank-2 tensors **A** and **B** we have:

$$A_{ij} = +A_{ji}$$
$$B_{ij} = -B_{ji}$$

where **A** is symmetric and **B** is anti-symmetric. Similar relations hold when **A** and **B** are contravariant.

2.75 Do we have symmetric/anti-symmetric scalars or vectors? If not, why?

No, because an exchange of indices is required in the definition of symmetric and anti-symmetric tensors and since scalars have no index and vectors have only one index it is impossible to have an exchange of indices and hence they cannot be symmetric or anti-symmetric.

2.76 Is it the case that any tensor of rank > 1 should be either symmetric or anti-symmetric?

No. There are many tensors which are neither symmetric nor anti-symmetric. For example, a tensor **A** with $A_{12} = 2$ and $A_{21} = 5$ is neither symmetric nor anti-symmetric.

2.77 Give an example, writing all the components in numbers or symbols, of a symmetric tensor of rank-2 in a 3D space. Do the same for an anti-symmetric tensor of the same rank.

Symmetric:

$$\begin{matrix} 1 & 22 & 3.1 \\ 22 & -4 & 1.9 \\ 3.1 & 1.9 & 6 \end{matrix}$$

Anti-symmetric:

$$\begin{matrix} 0 & a & b \\ -a & 0 & -c \\ -b & c & 0 \end{matrix}$$

2.78 Give, if possible, an example of a rank-2 tensor which is neither symmetric nor anti-symmetric assuming a 4D space.

Answer:

$$\begin{matrix} 1 & 2 & 3 & 4 \\ 5 & 6 & 7 & 8 \\ 9 & 10 & 11 & 12 \\ 13 & 14 & 15 & 16 \end{matrix}$$

2.79 Using the Index in the back of the book, gather all the terms related to the symmetric and anti-symmetric tensors including the symbols used in their notations.
Some of these terms are: anti-symmetry of tensor, contravariant tensor, covariant tensor, Kronecker δ, Levi-Civita tensor, metric tensor, parenthesis notation, partial anti-symmetry, partial symmetry, permutation of tensor, permutation tensor, skew-symmetric tensor, symmetry of tensor, totally anti-symmetric, totally symmetric, transposition of indices, unit tensor, zero tensor.

2.80 Is it true that any rank-2 tensor can be decomposed into a symmetric part and an anti-symmetric part? If so, write down the mathematical expressions representing these parts in terms of the original tensor. Is this also true for a general rank-n tensor?
Yes.
Mathematical expressions:

$$A_{ij} = A_{(ij)} + A_{[ij]} \qquad A_{(ij)} = \frac{1}{2}\left(A_{ij} + A_{ji}\right) \qquad A_{[ij]} = \frac{1}{2}\left(A_{ij} - A_{ji}\right)$$

where $A_{(ij)}$ is the symmetric part of A_{ij} and $A_{[ij]}$ is its anti-symmetric part.
A rank-n tensor can be symmetrized and anti-symmetrized by the relations given in the text, but the representation of the original tensor in terms of these symmetrized and anti-symmetrized tensors will not be as simple as in the case of rank-2.

2.81 What is the meaning of the round and square brackets which are used to contain indices in the indexed symbol of a tensor (e.g. $A_{(ij)}$ and $B^{[km]n}$)?
The round brackets mean that the tensor is symmetric with respect to the indices contained inside these brackets, while the square brackets mean that the tensor is anti-symmetric with respect to the indices inside the brackets.

2.82 Can the indices of symmetry/anti-symmetry be of different variance type?
No.

2.83 Is it possible that a rank-n ($n > 2$) tensor is symmetric/anti-symmetric with respect to some, but not all, of its indices? If so, give an example of a rank-3 tensor which is symmetric or anti-symmetric with respect to only two of its indices.

Yes.

A simple example is given in Figure 8 for a rank-3 tensor A_{ijk} in a 2D space which is symmetric in its first two indices (since $A_{121} = A_{211} = 1$ and $A_{122} = A_{212} = 2$) but it is neither symmetric in its last two indices (since $A_{112} = 0 \neq A_{121} = 1$ and $A_{212} = 2 \neq A_{221} = 0$) nor symmetric in its first and last indices (since $A_{112} = 0 \neq A_{211} = 1$ and $A_{122} = 2 \neq A_{221} = 0$). The tensor can be made partially anti-symmetric with respect to its first two indices if we change, for example, $A_{121} = 1$ to $A_{121} = -1$ and $A_{122} = 2$ to $A_{122} = -2$ (since $A_{121} = -A_{211} = -1$, $A_{122} = -A_{212} = -2$ and $A_{111} = A_{112} = A_{221} = A_{222} = 0$).

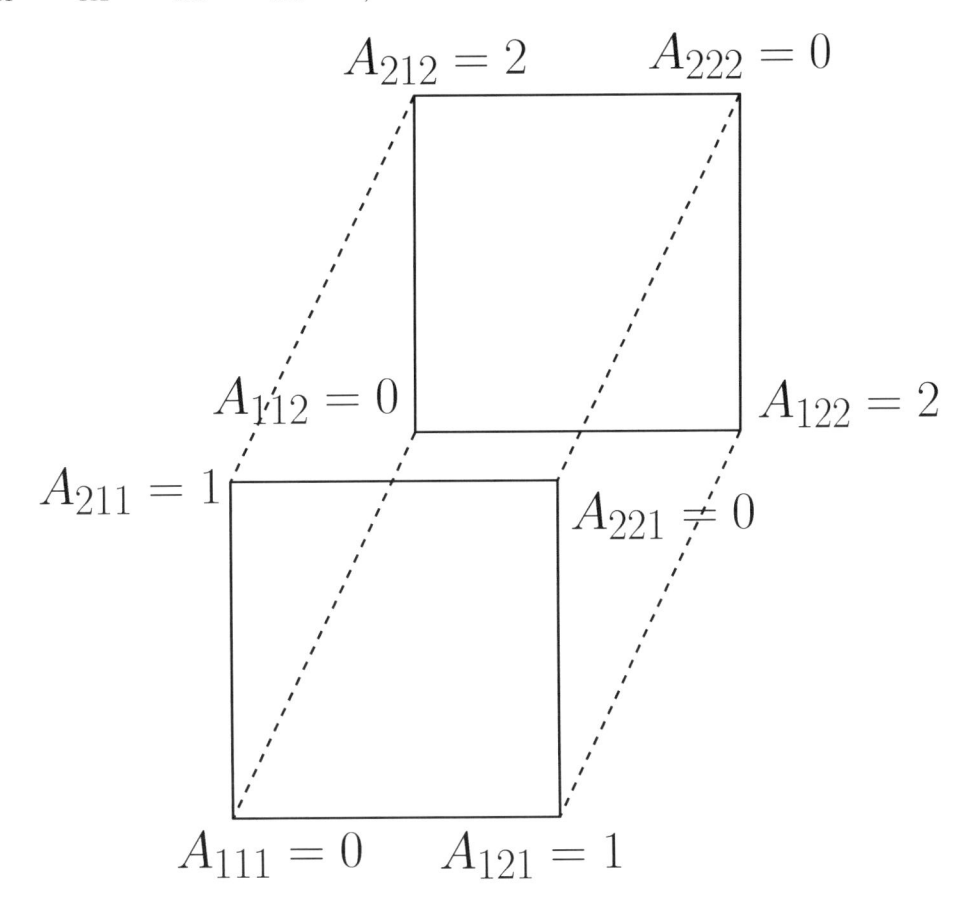

Figure 8: A partially symmetric rank-3 tensor A_{ijk} in a 2D space.

2.84 For a rank-3 covariant tensor A_{ijk}, how many possibilities of symmetry and anti-symmetry do we have? Consider in your answer total, as well as partial, symmetry and anti-symmetry. Is there another possibility (i.e. the tensor is neither symmetric nor anti-symmetric with respect to any pair of its indices)?

The tensor can be symmetric in the first and second indices only, or in the first and third indices only, or in the second and third indices only, or totally symmetric. If it is symmetric with respect to two different sets of its indices it is totally symmetric; for example if it is symmetric in its first and second indices and it is also symmetric in its second and third indices then we have:

$$A_{ijk} = A_{jik} \qquad \text{(first symmetry)}$$
$$= A_{jki} \qquad \text{(second symmetry)}$$
$$= A_{kji} \qquad \text{(first symmetry)}$$

and hence it is also symmetric in its first and third indices, so it is totally symmetric. Similarly, if it is symmetric in its first and second indices and in its first and third indices then we have:

$$A_{ijk} = A_{jik} \qquad \text{(first symmetry)}$$
$$= A_{kij} \qquad \text{(second symmetry)}$$
$$= A_{ikj} \qquad \text{(first symmetry)}$$

and hence it is also symmetric in its second and third indices, so it is totally symmetric. Likewise, if it is symmetric in its first and third indices and in its second and third indices then we have:

$$A_{ijk} = A_{kji} \qquad \text{(first symmetry)}$$
$$= A_{kij} \qquad \text{(second symmetry)}$$
$$= A_{jik} \qquad \text{(first symmetry)}$$

and hence it is also symmetric in its first and second indices, so it is totally symmetric.
The above also applies to anti-symmetry, i.e. the tensor can be anti-symmetric in the first and second indices only, or in the first and third indices only, or in the second and third indices only, or totally anti-symmetric. If it is anti-symmetric with respect to two different sets of its indices it is totally anti-symmetric; for example if it is anti-symmetric in its first and second indices and it is also anti-symmetric in its second

2   TENSORS   47

and third indices then we have:

$$A_{ijk} = -A_{jik} \quad \text{(first anti-symmetry)}$$
$$= +A_{jki} \quad \text{(second anti-symmetry)}$$
$$= -A_{kji} \quad \text{(first anti-symmetry)}$$

and hence it is also anti-symmetric in its first and third indices, so it is totally anti-symmetric. Similarly, if it is anti-symmetric in its first and second indices and in its first and third indices then we have:

$$A_{ijk} = -A_{jik} \quad \text{(first anti-symmetry)}$$
$$= +A_{kij} \quad \text{(second anti-symmetry)}$$
$$= -A_{ikj} \quad \text{(first anti-symmetry)}$$

and hence it is also anti-symmetric in its second and third indices, so it is totally anti-symmetric. Likewise, if it is anti-symmetric in its first and third indices and in its second and third indices then we have:

$$A_{ijk} = -A_{kji} \quad \text{(first anti-symmetry)}$$
$$= +A_{kij} \quad \text{(second anti-symmetry)}$$
$$= -A_{jik} \quad \text{(first anti-symmetry)}$$

and hence it is also anti-symmetric in its first and second indices, so it is totally anti-symmetric.

Yes, there is a possibility that the tensor is neither symmetric nor anti-symmetric with respect to any pair of its indices.

2.85 Can a tensor be symmetric with respect to some combinations of its indices and anti-symmetric with respect to the other combinations? If so, can you give a simple example of such a tensor?
Yes.
A trivial example is the zero tensor of any rank > 2 which is totally symmetric and totally anti-symmetric in its indices, hence we may consider its symmetry with respect to some combinations of its indices and its anti-symmetry with respect to the other combinations. However, if we want to find a tensor which is specifically symmetric

Table 1: Rank-4 tensor A_{ijkl} in 2D space.

i	j	k	l	A_{ijkl}
1	1	1	1	0
1	1	1	2	0
1	1	2	1	0
1	1	2	2	0
1	2	1	1	1
1	2	1	2	1
1	2	2	1	1
1	2	2	2	1
2	1	1	1	-1
2	1	1	2	-1
2	1	2	1	-1
2	1	2	2	-1
2	2	1	1	0
2	2	1	2	0
2	2	2	1	0
2	2	2	2	0

with respect to some combinations of its indices and specifically anti-symmetric with respect to the remaining combinations then we should look for another example. For a rank-3 tensor, we found in the previous exercise that if it is symmetric/anti-symmetric with respect to two different sets of its indices it is totally symmetric/anti-symmetric. So, we should search for such a tensor in higher ranks and this is beyond the scope of the book.

2.86 Repeat the previous exercise considering the additional possibility that the tensor is neither symmetric nor anti-symmetric with respect to another set of indices, i.e. it is symmetric, anti-symmetric and neither with respect to different sets of indices.
A simple example is shown in Table 1 where the rank-4 tensor A_{ijkl} in a 2D space is anti-symmetric in its first two indices and symmetric in its last two indices but, for example, it is neither symmetric nor anti-symmetric in its first and third indices.

2.87 **A** is a rank-3 totally symmetric tensor and **B** is a rank-3 totally anti-symmetric tensor. Write all the mathematical conditions that these tensors satisfy.
The main mathematical conditions are:

$$A_{ijk} = A_{jik} \qquad A_{ijk} = A_{ikj} \qquad A_{ijk} = A_{kji}$$

$$B_{ijk} = -B_{jik} \qquad B_{ijk} = -B_{ikj} \qquad B_{ijk} = -B_{kji}$$

In fact, any two of the three conditions should be enough as explained earlier in Exercise 2.84. Moreover, additional subsidiary conditions can be derived from these conditions (e.g. all components of **B** with two identical indices are zero).

2.88 Justify the following statement: "For a totally anti-symmetric tensor, non-zero entries can occur only when all the indices are different". Use mathematical, as well as descriptive, language in your answer.

This is because an exchange of two identical indices, which identifies the same entry, should change the sign due to the anti-symmetry and no number can be equal to its negation except zero. Therefore, if the tensor A_{ijk} is totally anti-symmetric then only its entries of the form A_{ijk}, where $i \ne j \ne k$, are not identically zero while all the other entries (i.e. all those of the from A_{iij}, A_{iji}, A_{jii} and A_{iii} where $i \ne j$ with no sum on i) vanish identically. In fact, this also applies to partially anti-symmetric tensors where the entries corresponding to identical anti-symmetric indices should vanish identically for the same reason. Hence, if A_{ijkl} is anti-symmetric in its first two indices and in its last two indices then only its entries with $i \ne j$ and $k \ne l$ do not vanish identically while all its entries with $i = j$ or $k = l$ (or both) are identically zero.

2.89 For a totally anti-symmetric tensor B_{ijk} in a 3D space, write all the elements of this tensor which are identically zero. Consider the possibility that it may be easier to find first the elements which are not identically zero, then exclude the rest.

This tensor has $3^3 = 27$ components. Because it is totally anti-symmetric, all the components with repetitive indices are identically zero and hence only the components with non-repetitive indices do not vanish identically. As there are only $3! = 6$ non-repetitive permutations of the three indices, we have only six non-identically vanishing components, these are:

$$B_{123}, B_{132}, B_{213}, B_{231}, B_{312}, B_{321}$$

while the remaining 21 components are identically zero. In fact, this question can be answered more easily by observing that the tensor B_{ijk} is like the rank-3 permutation tensor which is also totally anti-symmetric tensor in a 3D space and it has only six non-zero components.

Chapter 3
Tensor Operations

3.1 Give preliminary definitions of the following tensor operations: addition, multiplication by a scalar, tensor multiplication, contraction, inner product and permutation. Which of these operations involve a single tensor?
Addition: summing the corresponding components of two compatible (i.e. have the same dimension and identical indicial structure) tensors, e.g. $A_{ij} + B_{ij}$.
Multiplication by a scalar: multiplying each component of a non-scalar tensor by a given scalar, e.g. aB^{ijk}.
Tensor multiplication: multiplying each component of a non-scalar tensor of rank-m_1 by each component of another non-scalar tensor of rank-m_2 to produce a tensor of rank-$(m_1 + m_2)$ with $n^{m_1+m_2}$ components where n is the space dimension.
Contraction: converting two free indices in a tensor term to a dummy index, by unifying their symbols, followed by performing summation over this dummy index.
Inner product: an outer product operation (i.e. tensor multiplication) followed by contraction of some indices of the product.
Permutation: exchanging the indices of a tensor.
Permutation definitely involves a single tensor. Contraction can also involve a single tensor (e.g. A^i_i) although it can occur between two tensors as part of a product (e.g. $A_{ij}B^{jk}$), however the product may be considered as a single tensor. Multiplication of a tensor by a scalar my be considered as a single-tensor operation if the scalar is not considered as a tensor for some reason, i.e. either because "tensor" is used as opposite to scalar and vector or/and because the scalar is not a tensor in a general sense due to its failure to satisfy the principle of invariance.

3.2 Give typical examples of addition/subtraction for rank-n ($0 \leq n \leq 3$) tensors.
Answer:
$$a \pm b \qquad A_i \pm B_i \qquad A^{ij} \pm B^{ij} \qquad A^i_{jk} \pm B^i_{jk}$$

3.3 Is it possible to add two tensors of different ranks or different variance types? Is addition of tensors associative or commutative?
No. Addition (and subtraction) of tensors should satisfy the rules of free indices that

occur in the terms of tensor expressions and equalities (refer to the answer of question 2.23).

Yes, it is associative and commutative.

3.4 Discuss, in detail, the operation of multiplication of a tensor by a scalar and compare it to the operation of tensor multiplication. Can we regard multiplying two scalars as an example of multiplying a tensor by a scalar?

Multiplication of a tensor by a scalar means multiplying each component of a non-scalar tensor by a given scalar and therefore all the components of the non-scalar tensor are scaled by the scalar. Hence, the result of this operation is a tensor of the same rank and variance type as the original non-scalar tensor. The difference between this operation and the operation of tensor multiplication is that only one tensor in the former operation is non-scalar while both tensors in the latter operation are non-scalar.

Maybe, but this is very trivial case which may not be included in the common interpretation of "multiplying a tensor by a scalar" since the "tensor" here is supposed to be non-scalar according to the common understanding, as stated above. However, if "tensor" in "multiplication of a tensor by a scalar" is used in its general sense according to the principle of invariance and "scalar" is used in its generic sense as a single-component multiplication factor, then this extension may have a sensible interpretation.

3.5 What is the meaning of the term "outer product" and what are the other terms used to label this operation?

It means "tensor multiplication" as defined in the text and in the previous exercises. Outer product may also be called direct or exterior or dyadic product.

3.6 **C** is a tensor of rank-3 and **D** is a tensor of rank-2, what is the rank of their outer product **CD**? What is the rank of **CD** if it is subjected subsequently to a double contraction operation?

CD is of rank-5.

The double contraction operation will reduce the rank by 4 and hence the result is a rank-1 tensor.

3.7 **A** is a tensor of type (m, n) and **B** is a tensor of type (s, t), what is the type of their direct product **AB**?

The type of **AB** is: $(m + s, n + t)$.

3.8 Discuss the operations of dot and cross product of two vectors (see § Dot Product and Cross Product) from the perspective of the outer product operation of tensors.
The dot product of two vectors can be considered as an outer product operation which produces a rank-2 tensor followed by contraction which results in a scalar.
There is no obvious link between the operation of cross product of two vectors and the operation of outer product of tensors.

3.9 Collect from the Index all the terms related to the tensor operations of addition and permutation and classify these terms according to each operation giving a short definition of each.
Some of the relevant terms are:
Addition: associative (the order of operations can be changed), commutative (the order of operands can be changed), tensor operation (operation involving tensor or tensors).
Permutation: isomer (tensor obtained by permuting the indices of a given tensor), non-scalar tensor (tensor of rank > 0; here rank should be > 1), transpose of matrix (matrix obtained by exchanging the rows and columns of a given matrix), transposition of indices (exchanging neighboring indices), tensor operation (operation involving tensor or tensors).

3.10 Are the following two statements correct (make corrections if necessary)? "The outer multiplication of tensors is commutative but not distributive over sum of tensors" and "The outer multiplication of two tensors may produce a scalar".
Incorrect. Corrected: "The outer multiplication of tensors is not commutative but it is distributive over sum of tensors".
Incorrect. Corrected: "The outer multiplication of two tensors does not produce a scalar". However, the statement can be correct if we consider multiplication of two rank-0 tensors (i.e. "tensor" is used in its general sense that includes scalar and vector) as a special case of outer multiplication or as an extension to its definition.

3.11 What is contraction of tensor? How many free indices are consumed in a single contraction operation?
Contraction of tensor is the operation of converting two free indices of a tensor to a dummy index by unifying their label (e.g. changing A^i_j to A^i_i) followed by summation over this index. Hence, two free indices are consumed in a single contraction operation.

3.12 Is it possible that the contracted indices are of the same variance type? If so, what is

3 TENSOR OPERATIONS

the condition that should be satisfied for this to happen?

Not in general, although this is possible in orthonormal Cartesian systems where the variance type is irrelevant and hence there is no difference between lower and upper indices.

3.13 **A** is a tensor of type (m, n) where $m, n > 1$, what is its type after two contraction operations assuming a general coordinate system?

The type will be $(m - 2, n - 2)$ because each contraction operation consumes one contravariant index and one covariant index.

3.14 Does the contraction operation change the weight of a relative tensor?

No. As stated in the text, the application of a contraction operation on a relative tensor produces a relative tensor of the same weight as the original tensor. This also applies to absolute tensors since the weight (which is zero in this case) does not change by contraction.

3.15 Explain how the operation of multiplication of two matrices, as defined in linear algebra, involves a contraction operation. What is the rank of each matrix and what is the rank of the product? Is this consistent with the rule of reduction of rank by contraction?

The multiplication of two $n \times n$ matrices to produce another $n \times n$ matrix involves a contraction operation because matrix multiplication can be regarded as an outer multiplication that produces a rank-4 tensor followed by contracting the middle indices in the product (e.g. $A_{ij} B^{jk}$).

The rank of each one of the matrices involved in matrix multiplication, as well as the rank of the product, is 2.

Yes, this is consistent with the rule of reduction of rank by contraction because the outer product of two rank-2 tensors is a rank-4 tensor and the contraction operation on the product consumes two free indices, hence reducing its rank by 2 and making it a rank-2 tensor as well.

3.16 Explain, in detail, the operation of inner product of two tensors and how it is related to the operations of contraction and outer product of tensors.

Inner multiplication of two tensors is a composite operation that consists of an outer multiplication followed by contraction of two indices of the product. Hence, the inner product of a tensor of type (m_1, n_1) with a tensor of type (m_2, n_2) is a tensor of type

$(m_1 + m_2 - 1, n_1 + n_2 - 1)$. Accordingly, the inner product of a rank-p tensor with a rank-q tensor is a rank-$(p + q - 2)$ tensor.

3.17 What is the rank and type of a tensor resulting from an inner product operation of a tensor of type (m, n) with a tensor of type (s, t)? How many possibilities do we have for this inner product considering the different possibilities of the embedded contraction operation?
The rank is $(m + s + n + t - 2)$ and the type is $(m + s - 1, n + t - 1)$.
There are $(m + s) \times (n + t)$ possibilities for this inner product. These possibilities represent all the possible combinations of contracting one upper index with one lower index in the product.

3.18 Give an example of a commutative inner product of two tensors and another example of a non-commutative inner product.
Commutative: $A_i B^i$.
Non-commutative: $A_{ij} B^{jk}$.

3.19 Is the inner product operation distributive over algebraic addition of tensors?
Yes.

3.20 Give an example from matrix algebra of inner product of tensors explaining in detail how the two are related.
An example from matrix algebra of inner product of tensors is the multiplication of an $n \times n$ matrix $\mathbf{A}$, representing a rank-2 tensor in nD space, by an n-component vector $\mathbf{b}$, representing a rank-1 tensor in nD space, to produce an n-component vector $\mathbf{c}$. This can be represented symbolically by the following operation:

$$[\mathbf{Ab}]_{ij}^{k} = A_{ij} b^k \quad \xrightarrow{\text{contraction on } jk} \quad c_i = [\mathbf{A} \cdot \mathbf{b}]_i = A_{ij} b^j$$

where the equality on the left represents the outer product that associates this operation while the equality on the right represents the final product of this inner multiplication which is obtained following the seen contraction of two indices one from the rank-2 tensor and one from the vector. The boldfaced expressions are in symbolic notation of tensors not in matrix or vector notation.

3.21 Discuss specialized types of inner product operations that involve more than one contraction operation focusing in particular on the operations $\mathbf{A} : \mathbf{B}$ and $\mathbf{A} \cdot\cdot\, \mathbf{B}$ where

A and **B** are two tensors of rank > 1.

The specialized inner product operations are similarly defined as the ordinary inner product operations but they involve more than one contraction operation and hence they can only apply to tensors or tensor products of rank > 3 since they require at least 4 free indices to consume in the multiple contraction operations. The most common of these operations are the double contraction inner multiplications which are symbolized by vertically- or horizontally-aligned two dots (i.e. : and $\cdot\cdot$). For two rank-2 tensors **A** and **B**, these operations are defined as follows:

$$\mathbf{A} : \mathbf{B} = A_{ij}B^{ij} \qquad\qquad \mathbf{A} \cdot\cdot\, \mathbf{B} = A_{ij}B^{ji}$$

where in the first operation the double contraction takes place between the corresponding indices of the two tensors, while in the second operation the double contraction takes place between the first/second index of the first tensor with the second/first index of the second tensor. The above double contraction inner product operations are defined by some authors opposite to the above definitions (i.e. the other way around).

3.22 A double inner product operation is conducted on a tensor of type $(1,1)$ with a tensor of type $(1,2)$. How many possibilities do we have for this operation? What is the rank and type of the resulting tensor? Is it covariant, contravariant or mixed?

The outer product of these tensors is of type $(2,3)$ and hence there is only one possibility for the choice of the two upper indices needed by the double contraction operation but there are three possibilities for the two lower indices needed by the double contraction operation, i.e. index 1 with index 2, index 1 with index 3 and index 2 with index 3. Therefore, there are three possibilities for this operation.

The rank of the resulting tensor is 1 and the type is $(0,1)$.

It is covariant.

3.23 Gather from the Index all the terms that refer to notations used in the operations of inner and outer product of tensors.

Direct multiplication symbol ($\otimes$), dot notation ($\cdot$), double dot notation (: and $\cdot\cdot$).

3.24 Assess the following statement considering the two meanings of the word "tensor" related to the rank: "Inner product operation of two tensors does not necessarily produce a tensor". Can this statement be correct in a sense and wrong in another?

Inner product operation of two tensors does not necessarily produce a tensor if "tensor" is used as opposite to scalar and vector because inner product operation can produce scalar or vector and hence it does not necessarily produce a tensor in this sense. However, if "tensor" is used in its general sense based on its transformation properties and hence it can be of rank-0 or rank-1, then inner product operation of two tensors does necessarily produce a tensor in this sense even when the resulting tensor is of rank-0 or rank-1.

3.25 What is the operation of tensor permutation and how it is related to the operation of transposition of matrices?

The operation of tensor permutation is the exchange of tensor indices. Transposition of matrices is an instance of tensor permutation (when these matrices represent tensors) because transposition of matrices means the exchange of their two indices which refer to the rows and columns.

3.26 Is it possible to permute scalars or vectors and why?

No, because permutation requires 2 indices at least and scalars have no index and vectors have only one index and hence permutation of indices cannot take place.

3.27 What is the meaning of the term "isomers"?

Isomer of a tensor $\mathbf{A}$ is a tensor obtained by permuting the indices of $\mathbf{A}$. For example, A^i_{kj} is an isomer of the tensor A^i_{jk}.

3.28 Describe in detail the quotient rule and how it is used as a test for tensors.

The quotient rule of tensors (which is different from the quotient rule of differentiation) is a relatively easy test that can be conducted on a mathematical object, which is suspected to be a tensor, to verify if it is really a tensor or not. According to the quotient rule of tensors, if the inner product of a suspected tensor by a known tensor is a tensor then the suspect is a tensor. For example, if it is not known if $\mathbf{A}$ is a tensor or not but it is known that $\mathbf{B}$ and $\mathbf{C}$ are tensors; moreover it is known that the following relation holds true in all rotated coordinate systems:

$$C^{pq...m}{}_{ij...n} = A^{pq...k...m} B_{ij...k...n}$$

then $\mathbf{A}$ is a tensor.

3.29 Why the quotient rule is used instead of the standard transformation equations of tensors?

Because using the quotient rule is generally more convenient and requires less work than employing the first principles by direct application of the transformation rules. Moreover, the quotient rule may not require actual work when we know, from our past experience, that a relation like the one seen in the previous question is already satisfied by our tensors so all we need from the quotient rule is to draw the conclusion that the suspected tensor is a tensor indeed.

Chapter 4

δ and ϵ Tensors

4.1 What "numerical tensor" means in connection with the Kronecker δ and the permutation ϵ tensors?

It means their components are numbers which are $0, 1$ for the Kronecker δ tensor and $0, 1, -1$ for the permutation tensor.

4.2 State all the names used to label the Kronecker and permutation tensors.
Kronecker: Kronecker delta tensor, Kronecker symbol, unit tensor, identity tensor, identity matrix.
Permutation: Levi-Civita tensor, permutation tensor, epsilon tensor, anti-symmetric tensor, alternating tensor (or symbol).

4.3 What is the meaning of "conserved under coordinate transformations" in relation to the Kronecker and permutation tensors?

It means the values of the tensor components do not change following a coordinate transformation. For example, $\delta_1^1 = 1$ and $\epsilon^{112} = 0$ before and after transformation.

4.4 State the mathematical definition of the Kronecker δ tensor.
Answer:
$$\delta_{ij} = \begin{cases} 1 & (i = j) \\ 0 & (i \neq j) \end{cases} \qquad (i, j = 1, 2, \ldots n)$$
where n is the space dimension. This definition similarly applies to the contravariant and mixed forms of this tensor.

4.5 What is the rank of the Kronecker δ tensor in an nD space?

The rank is 2 because it has 2 free indices regardless of the space dimension.

4.6 Write down the matrix representing the Kronecker δ tensor in a 3D space.
Answer:
$$\begin{bmatrix} 1 & 0 & 0 \\ 0 & 1 & 0 \\ 0 & 0 & 1 \end{bmatrix}$$

4.7 Is there any difference between the components of the covariant, contravariant and mixed types of the Kronecker δ tensor?

No. For example, $\delta_{22} = \delta^{22} = \delta^2_2 = 1$ and $\delta_{21} = \delta^{21} = \delta^2_1 = \delta^1_2 = 0$.

4.8 Explain how the Kronecker δ acts as an index replacement operator giving an example in a mathematical form.

When one index of the Kronecker delta tensor is a dummy index by having the same label as the label of an index in another tensor in its own term, the shared index in the other tensor is replaced by the other index of the Kronecker delta while keeping its position as upper or lower. Examples:

$$\delta^j_i A_j = A_i \qquad \delta^i_j A^j = A^i$$

In such cases the Kronecker delta is described as an index replacement or substitution operator.

4.9 How many mathematical definitions of the rank-n permutation tensor we have? State one of these definitions explaining all the symbols involved.

We have three: inductive definition, analytic definition and a definition based on the generalized Kronecker delta tensor. The latter is given by:

$$\epsilon_{i_1 \ldots i_n} = \delta^{1 \ldots n}_{i_1 \ldots i_n} \qquad \epsilon^{i_1 \ldots i_n} = \delta^{i_1 \ldots i_n}_{1 \ldots n}$$

where the indexed ϵ is the permutation tensor in its covariant and contravariant forms, the indexed δ is the generalized Kronecker delta, $i_1 \ldots i_n$ are indices, and n is the space dimension.

4.10 What is the rank of the permutation tensor in an nD space?

It is n.

4.11 Make a graphical illustration of the array representing the rank-2 and rank-3 permutation tensors.

The graphical illustration of rank-2 permutation tensor is seen in Figure 9, while the graphical illustration of rank-3 permutation tensor is seen in Figure 10. In these figures, the black nodes represent the 0 components, the blue squares represent the 1 components and the red triangles represent the -1 components.

4.12 Is there any difference between the components of the covariant and contravariant

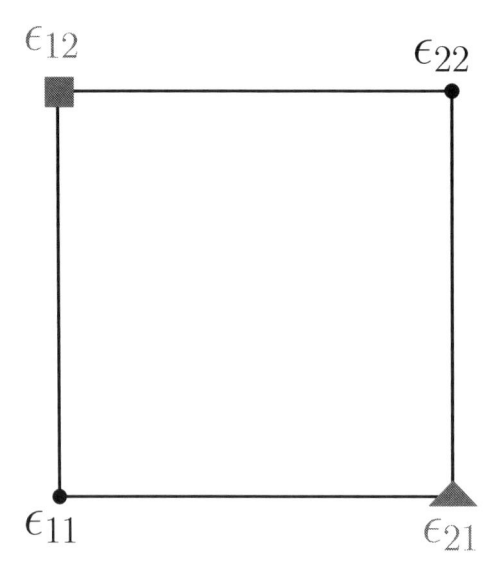

Figure 9: Graphical illustration of the rank-2 permutation tensor ϵ_{ij}.

types of the permutation tensor?

No. For example, $\epsilon_{132} = -1$ and $\epsilon^{132} = -1$.

4.13 How the covariant and contravariant types of the permutation tensor are related to the concept of relative tensor?

The covariant type of the permutation tensor is a relative tensor of weight $w = -1$, while the contravariant type of the permutation tensor is a relative tensor of weight $w = +1$.

4.14 State the distinctive properties of the permutation tensor.

The main properties of the permutation tensor are:

(a) It is numeric tensor and hence the value of its components are: -1, 1 and 0 in all coordinate systems.

(b) The value of any particular component (e.g. ϵ_{312}) of this tensor is the same in any coordinate system and hence it is constant tensor in this sense.

(c) It is relative tensor of weight -1 for its covariant form and $+1$ for its contravariant form.

(d) It is isotropic tensor since its components are conserved under proper transformations.

(e) It is totally anti-symmetric in each pair of its indices, i.e. it changes sign on swapping any two of its indices.

(f) It is pseudo tensor since it acquires a minus sign under improper orthogonal trans-

4 δ AND ε TENSORS

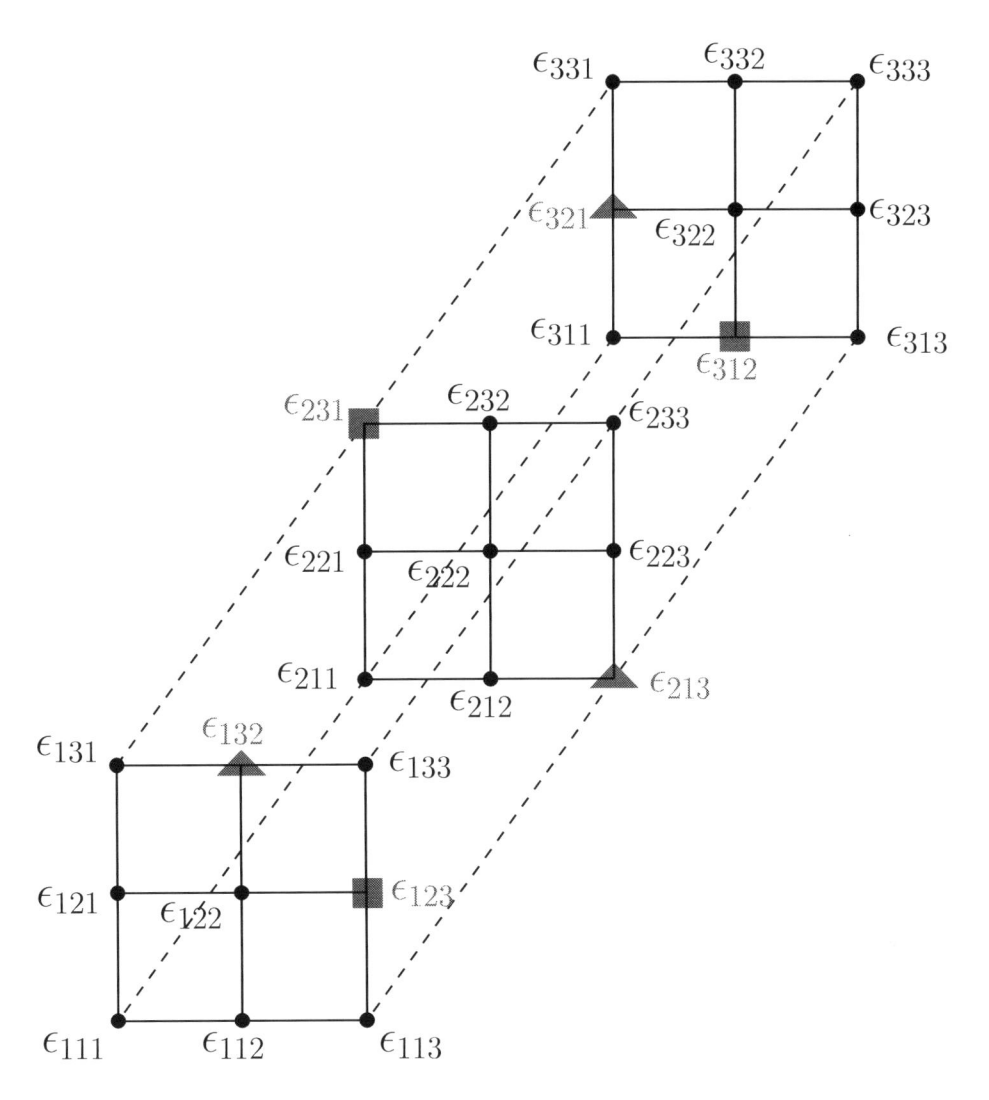

Figure 10: Graphical illustration of the rank-3 permutation tensor ϵ_{ijk}.

formation of coordinates.

(g) The permutation tensor of any rank has only one independent non-vanishing component because all the non-zero components of this tensor are of unity magnitude.

(h) The rank-n permutation tensor possesses $n!$ non-zero components which is the number of the non-repetitive permutations of its indices.

(i) The rank and the dimension of the permutation tensor are identical and hence in an nD space it has n^n components.

4.15 How many entries the rank-3 permutation tensor has? How many non-zero entries it has? How many non-zero independent entries it has?

The rank and dimension of the permutation tensor are identical and hence the rank-3 permutation tensor has $3^3 = 27$ components (see Figure 10).

The number of non-repetitive permutations of the three indices is $n! = 3! = 6$ and hence it has 6 non-zero entries.

It has only one non-zero independent entry because all the non-zero entries are ± 1 and hence when they differ, they differ only in sign.

4.16 Is the permutation tensor true or pseudo and why?

It is pseudo tensor because it acquires a minus sign under improper orthogonal transformation of coordinates which changes system handedness and hence the parity of the permutations of indices. For example, on exchanging the second and third coordinates of a 3D right handed Cartesian system, ϵ_{123} will become ϵ_{132} since the order of the labels of the two coordinates will change by this improper transformation.

4.17 State, in words, the cyclic property of the even and odd non-repetitive permutations of the rank-3 permutation tensor with a simple sketch to illustrate this property.

The cyclic property means that if the last index is put first (or the other way around) the parity (even or odd) of the permutation is preserved. Hence, the three even permutations are obtained from each other by this cyclic exchange and similarly for the odd permutations. This is demonstrated graphically in Figure 11.

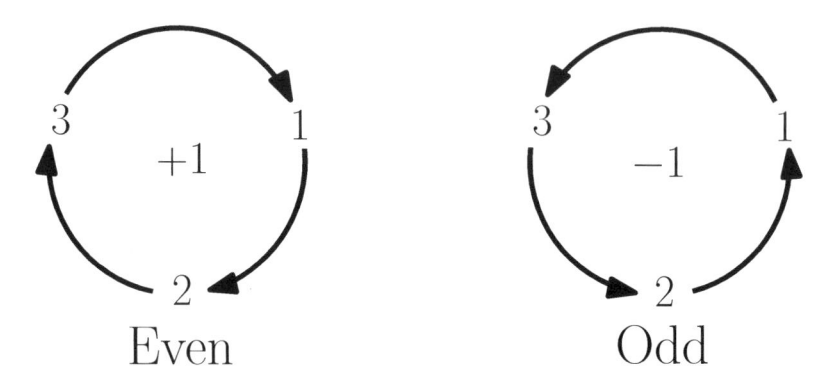

Figure 11: The cyclic property of the even and odd permutations of the indices of the rank-3 permutation tensor.

4.18 From the Index, find all the terms which are common to the Kronecker and permutation tensors.

Contracted epsilon identity, contravariant tensor, covariant tensor, epsilon-delta identity, isotropic tensor, Levi-Civita identity, numerical tensor, tensor.

4.19 Correct the following equations:

$$\delta_{ij}A_j = A_j \qquad \delta_{ij}\delta_{jk} = \delta_{jk} \qquad \delta_{ij}\delta_{jk}\delta_{ki} = n! \qquad x_{i,j} = \delta_{ii}$$

Corrected:

$$\delta_{ij}A_j = A_i \qquad \delta_{ij}\delta_{jk} = \delta_{ik} \qquad \delta_{ij}\delta_{jk}\delta_{ki} = n \qquad x_{i,j} = \delta_{ij}$$

4.20 In what type of coordinate system the following equation applies?

$$\partial_i x_j = \partial_j x_i$$

Answer: In orthonormal Cartesian coordinate systems.

4.21 Complete the following equation assuming a 4D space:

$$\partial_i x_i = ?$$

Completed:
$$\partial_i x_i = \delta_{ii} = n = 4$$

4.22 Complete the following equations where the indexed **e** are orthonormal basis vectors of a particular coordinate system:

$$\mathbf{e}_i \cdot \mathbf{e}_j = ? \qquad\qquad \mathbf{e}_i\mathbf{e}_j : \mathbf{e}_k\mathbf{e}_l = ?$$

Completed:

$$\mathbf{e}_i \cdot \mathbf{e}_j = \delta_{ij} \qquad\qquad \mathbf{e}_i\mathbf{e}_j : \mathbf{e}_k\mathbf{e}_l = \delta_{ik}\delta_{jl}$$

4.23 Write down the equations representing the cyclic order of the rank-3 permutation tensor. What is the conclusion from these equations with regard to the symmetry or anti-symmetry of this tensor and the number of its independent non-zero components?
Cyclic order equations:

$$\epsilon_{ijk} = \epsilon_{kij} = \epsilon_{jki} = -\epsilon_{ikj} = -\epsilon_{jik} = -\epsilon_{kji}$$

The conclusion is that this tensor is totally anti-symmetric because an exchange of any two indices leads to reversal of sign. Another conclusion is that this tensor has only one independent non-vanishing component because all the six non-vanishing components are equal except (possibly) in sign.

4.24 Write the analytical expressions of the rank-3 and rank-4 permutation tensors.
The values of the components of the rank-3 permutation tensor are given analytically by:
$$\epsilon_{ijk} = \epsilon^{ijk} = \frac{1}{2}(j-i)(k-i)(k-j)$$
where each one of the indices i, j, k ranges over $1, 2, 3$.
The values of the components of the rank-4 permutation tensor are given analytically by:
$$\epsilon_{ijkl} = \epsilon^{ijkl} = \frac{1}{12}(j-i)(k-i)(l-i)(k-j)(l-j)(l-k)$$
where each one of the indices i, j, k, l ranges over $1, 2, 3, 4$.

4.25 Collect from the Index all the terms from matrix algebra which have connection to the Kronecker δ.
Diagonal matrix, identity matrix, invertible matrix, inverse of matrix, non-singular matrix, square matrix, unit matrix, unity matrix.

4.26 Correct, if necessary, the following equations:
$$\epsilon_{i_1 \cdots i_n} \epsilon_{i_1 \cdots i_n} = n \quad \epsilon_{ijk} C_j C_k = 0 \quad \epsilon_{ijk} D_{jk} = 0 \quad \mathbf{e}_i \times \mathbf{e}_j = \epsilon_{ijk} \mathbf{e}_j \quad (\mathbf{e}_i \times \mathbf{e}_j) \cdot \mathbf{e}_k = \epsilon_{ijk}$$

where $\mathbf{C}$ is a vector, $\mathbf{D}$ is a symmetric rank-2 tensor, and the indexed $\mathbf{e}$ are orthonormal basis vectors in a 3D space with a right-handed coordinate system.
The second, third and fifth equations are correct while the first and fourth equations should be:
$$\epsilon_{i_1 \cdots i_n} \epsilon_{i_1 \cdots i_n} = n! \qquad \mathbf{e}_i \times \mathbf{e}_j = \epsilon_{ijk} \mathbf{e}_k$$

4.27 What is wrong with the following equations?
$$\epsilon_{ijk} \delta_{1i} \delta_{2j} \delta_{3k} = -1 \qquad \epsilon_{ij} \epsilon_{kl} = \delta_{ik} \delta_{jl} + \delta_{il} \delta_{jk} \qquad \epsilon_{il} \epsilon_{kl} = \delta_{il} \qquad \epsilon_{ij} \epsilon_{ij} = 3!$$

All these equations are wrong. They should be corrected as follows:

$$\epsilon_{ijk}\delta_{1i}\delta_{2j}\delta_{3k} = +1 \qquad \epsilon_{ij}\epsilon_{kl} = \delta_{ik}\delta_{jl} - \delta_{il}\delta_{jk} \qquad \epsilon_{il}\epsilon_{kl} = \delta_{ik} \qquad \epsilon_{ij}\epsilon_{ij} = 2!$$

4.28 Write the following in their determinantal form describing the general pattern of the relation between the indices of ϵ and δ and the indices of the rows and columns of the determinant:

$$\epsilon_{ijk}\epsilon_{lmk} \qquad \epsilon_{ijk}\epsilon_{lmn} \qquad \epsilon_{i_1\cdots i_n}\epsilon_{j_1\cdots j_n} \qquad \delta^{i_1\ldots i_n}_{j_1\ldots j_n}$$

Answer:

$$\epsilon_{ijk}\epsilon_{lmk} = \begin{vmatrix} \delta_{il} & \delta_{im} \\ \delta_{jl} & \delta_{jm} \end{vmatrix}$$

Pattern: ignoring the repeated index k, the indices of the first ϵ index the rows while the indices of the second ϵ index the columns.

$$\epsilon_{ijk}\epsilon_{lmn} = \begin{vmatrix} \delta_{il} & \delta_{im} & \delta_{in} \\ \delta_{jl} & \delta_{jm} & \delta_{jn} \\ \delta_{kl} & \delta_{km} & \delta_{kn} \end{vmatrix}$$

Pattern: the indices of the first ϵ index the rows while the indices of the second ϵ index the columns.

$$\epsilon_{i_1\cdots i_n}\epsilon_{j_1\cdots j_n} = \begin{vmatrix} \delta_{i_1 j_1} & \delta_{i_1 j_2} & \cdots & \delta_{i_1 j_n} \\ \delta_{i_2 j_1} & \delta_{i_2 j_2} & \cdots & \delta_{i_2 j_n} \\ \vdots & \vdots & \ddots & \vdots \\ \delta_{i_n j_1} & \delta_{i_n j_2} & \cdots & \delta_{i_n j_n} \end{vmatrix}$$

Pattern: the indices of the first ϵ index the rows while the indices of the second ϵ index the columns.

$$\delta^{i_1\ldots i_n}_{j_1\ldots j_n} = \begin{vmatrix} \delta^{i_1}_{j_1} & \delta^{i_1}_{j_2} & \cdots & \delta^{i_1}_{j_n} \\ \delta^{i_2}_{j_1} & \delta^{i_2}_{j_2} & \cdots & \delta^{i_2}_{j_n} \\ \vdots & \vdots & \ddots & \vdots \\ \delta^{i_n}_{j_1} & \delta^{i_n}_{j_2} & \cdots & \delta^{i_n}_{j_n} \end{vmatrix}$$

Pattern: the upper indices of the generalized Kronecker delta provide the upper indices of the ordinary Kronecker deltas in the determinant array by indexing the rows of the determinant, while the lower indices of the generalized Kronecker delta provide the lower indices of the ordinary Kronecker deltas in the determinant array by indexing the columns of the determinant.

4.29 Give two mnemonic devices used to memorize the widely used epsilon-delta identity and make a simple graphic illustration for one of these.

One mnemonic device is the determinantal form of this identity where the expanded form, if needed, can be easily obtained from the determinant which can be easily built following the simple pattern of indices, as explained in the answer of the previous question. Another mnemonic device is shown in Figure 12 where the first and second bars (i.e. ||) represent respectively the first two indices and the second two indices of the two epsilons which are used as indices for the first and second deltas on the right, while the cross (i.e. ×) represents the first index of the first epsilon with the second index of the second epsilon which are used as indices for the third delta and the second index of the first epsilon with the first index of the second epsilon which are used as indices for the fourth delta on the right.

$$\epsilon_{ijk}$$

$$\epsilon_{lmk}$$

$$|\,| - \times$$

$$\delta_{il}\delta_{jm} - \delta_{im}\delta_{jl}$$

Figure 12: Mnemonic device for the epsilon-delta identity.

4.30 Correct, if necessary, the following equations:

$$\epsilon_{rst}\epsilon_{rst} = 3! \qquad \epsilon_{pst}\epsilon_{qst} = 2\delta_{pq} \qquad \epsilon_{rst}\delta_{rt} = \epsilon_{rst}\delta_{st}$$

All these equations are correct.

4 δ AND ε TENSORS

4.31 State the mathematical definition of the generalized Kronecker delta $\delta^{i_1...i_n}_{j_1...j_n}$.
Answer:
$$\delta^{i_1...i_n}_{j_1...j_n} = \begin{cases} 1 & (j_1...j_n \text{ is even permutation of } i_1...i_n) \\ -1 & (j_1...j_n \text{ is odd permutation of } i_1...i_n) \\ 0 & (\text{repeated } j\text{'s}) \end{cases}$$

4.32 Write each one of $\epsilon_{i_1...i_n}$ and $\epsilon^{i_1...i_n}$ in terms of the generalized Kronecker δ.
Answer:
$$\epsilon_{i_1...i_n} = \delta^{1...n}_{i_1...i_n} \qquad\qquad \epsilon^{i_1...i_n} = \delta^{i_1...i_n}_{1...n}$$

4.33 Make a survey, based on the Index, about the general mathematical terms used in the operations conducted by using the Kronecker and permutation tensors.
Some of these terms are: cross product, curl operation, index replacement operator, scalar triple product, substitution operator, vector triple product.

4.34 Write the mathematical relation that links the contravariant permutation tensor, the covariant permutation tensor, and the generalized Kronecker delta.
Answer:
$$\epsilon^{i_1...i_n} \epsilon_{j_1...j_n} = \delta^{i_1...i_n}_{j_1...j_n}$$

4.35 State the widely used epsilon-delta identity in terms of the generalized and ordinary Kronecker deltas.
Answer:
$$\delta^{ijk}_{lmk} = \delta^{ij}_{lm} = \delta^i_l \delta^j_m - \delta^i_m \delta^j_l$$

Chapter 5

Applications of Tensor Notation and Techniques

5.1 Write, in tensor notation, the mathematical expression for the trace and determinant of an $n \times n$ matrix and the inverse of a 3×3 matrix.
Answer:

$$\begin{aligned} \operatorname{tr}(\mathbf{A}) &= A_{ii} \\ \det(\mathbf{A}) &= \frac{1}{n!} \epsilon_{i_1 \cdots i_n} \epsilon_{j_1 \cdots j_n} A_{i_1 j_1} \ldots A_{i_n j_n} \\ \left[\mathbf{A}^{-1}\right]_{ij} &= \frac{1}{2 \det(\mathbf{A})} \epsilon_{jmn} \epsilon_{ipq} A_{mp} A_{nq} \end{aligned}$$

5.2 Repeat the previous exercise for the multiplication of a matrix by a vector and the multiplication of two $n \times n$ matrices.
Answer:

$$\begin{aligned} \left[\mathbf{A}\mathbf{b}\right]_i &= A_{ij} b_j \\ \left[\mathbf{A}\mathbf{B}\right]_{ik} &= A_{ij} B_{jk} \end{aligned}$$

5.3 Define mathematically the dot and cross product operations of two vectors using tensor notation.
Answer:

$$\begin{aligned} \mathbf{A} \cdot \mathbf{B} &= A_i B_i \\ \left[\mathbf{A} \times \mathbf{B}\right]_i &= \epsilon_{ijk} A_j B_k \end{aligned}$$

5.4 Repeat the previous exercise for the scalar triple product and vector triple product operations of three vectors.

5 APPLICATIONS OF TENSOR NOTATION AND TECHNIQUES

Answer:

$$\mathbf{A} \cdot (\mathbf{B} \times \mathbf{C}) = \epsilon_{ijk} A_i B_j C_k$$
$$[\mathbf{A} \times (\mathbf{B} \times \mathbf{C})]_i = \epsilon_{ijk}\epsilon_{klm} A_j B_l C_m$$

5.5 Define mathematically, using tensor notation, the three scalar invariants of a rank-2 tensor: I, II and III.
Answer:

$$I = A_{ii}$$
$$II = A_{ij}A_{ji}$$
$$III = A_{ij}A_{jk}A_{ki}$$

5.6 Express the three scalar invariants I, II and III in terms of the other three invariants I_1, I_2 and I_3 and vice versa.
Answer:

$$I = I_1$$
$$II = I_1^2 - 2I_2$$
$$III = I_1^3 - 3I_1 I_2 + 3I_3$$

$$I_1 = I$$
$$I_2 = \frac{1}{2}\left(I^2 - II\right)$$
$$I_3 = \frac{1}{3!}\left(I^3 - 3I\, II + 2III\right)$$

5.7 Explain why the three invariants I_1, I_2 and I_3 are scalars using in your argument the fact that the three main invariants I, II and III are traces?
The three main invariants I, II and III are traces and hence they are scalars. Because the three subsidiary invariants I_1, I_2 and I_3 can be expressed as algebraic combinations of the three main invariants, as seen in the answer of the previous question, they should also be scalars.

5.8 Gather six terms from the Index about the scalar invariants of tensors.

Determinant, invariant, length of vector, scalar, tensor, trace.

5.9 Justify, giving a detailed explanation, the following statement: "If a rank-2 tensor is invertible in a particular coordinate system it is invertible in all other coordinate systems, and if it is singular in a particular coordinate system it is singular in all other coordinate systems". Use in your explanation the fact that the determinant is invariant under admissible coordinate transformations.

The necessary and sufficient condition for a matrix to be invertible is that its determinant is not zero. Now, because the determinant of a rank-2 tensor is invariant then if the determinant vanishes in one coordinate system it will vanish in all transformed coordinate systems, and if does not vanish in one coordinate system it will not vanish in any transformed coordinate system. This means that the determinant of a given rank-2 tensor is either zero in all systems or not zero in all systems, i.e. it is either singular in all systems or it is invertible in all systems. Consequently, "if a rank-2 tensor is invertible in a particular coordinate system it is invertible in all other coordinate systems, and if it is singular in a particular coordinate system it is singular in all other coordinate systems" which is the given statement.

5.10 What are the ten joint invariants between two rank-2 tensors?

They are: $\text{tr}(\mathbf{A})$, $\text{tr}(\mathbf{B})$, $\text{tr}(\mathbf{A}^2)$, $\text{tr}(\mathbf{B}^2)$, $\text{tr}(\mathbf{A}^3)$, $\text{tr}(\mathbf{B}^3)$, $\text{tr}(\mathbf{A} \cdot \mathbf{B})$, $\text{tr}(\mathbf{A}^2 \cdot \mathbf{B})$, $\text{tr}(\mathbf{A} \cdot \mathbf{B}^2)$ and $\text{tr}(\mathbf{A}^2 \cdot \mathbf{B}^2)$.

5.11 Provide a concise mathematical definition of the nabla differential operator ∇ in Cartesian coordinate systems using tensor notation.

Answer:
$$\nabla_i = \frac{\partial}{\partial x_i} \qquad \text{or} \qquad \nabla = \mathbf{e}_i \frac{\partial}{\partial x_i}$$

5.12 What is the rank and variance type of the gradient of a differentiable scalar field in general curvilinear coordinate systems?

The rank is 1 and the variance type is covariant.

5.13 State, in tensor notation, the mathematical expression for the gradient of a differentiable scalar field in a Cartesian system.

Answer:
$$[\nabla f]_i = \nabla_i f = \frac{\partial f}{\partial x_i} = \partial_i f = f_{,i}$$

5 APPLICATIONS OF TENSOR NOTATION AND TECHNIQUES   71

5.14 What is the gradient of the following scalar functions of position f, g and h where x_1, x_2 and x_3 are the Cartesian coordinates and a, b and c are constants?

$$f = 1.3x_1 - 2.6ex_2 + 19.8x_3 \qquad g = ax_3 + be^{x_2} \qquad h = a(x_1)^3 - \sin x_3 + c(x_3)^2$$

Answer:

$$\begin{aligned} \nabla f &= (1.3,\ -2.6e,\ 19.8) \\ \nabla g &= (0,\ be^{x_2},\ a) \\ \nabla h &= \left(3a(x_1)^2,\ 0,\ -\cos x_3 + 2cx_3\right) \end{aligned}$$

5.15 State, in tensor notation, the mathematical expression for the gradient of a differentiable vector field in a Cartesian system.
Answer:
$$[\nabla \mathbf{A}]_{ij} = \partial_i A_j$$

5.16 What is the gradient of the following vector where x_1, x_2 and x_3 are the Cartesian coordinates?
$$\mathbf{V} = (2x_1 - 1.2x_2,\ x_1 + x_3,\ x_2 x_3)$$

What is the rank of this gradient?
Answer:

$$\begin{aligned} \nabla[\mathbf{V}]_1 &= (2,\ -1.2,\ 0) \\ \nabla[\mathbf{V}]_2 &= (1,\ 0,\ 1) \\ \nabla[\mathbf{V}]_3 &= (0,\ x_3,\ x_2) \end{aligned}$$

The rank is 2.

5.17 Explain, in detail, why the divergence of a vector is invariant.
The divergence operation is achieved by taking the gradient of the vector (which produces a rank-2 tensor) followed by taking the trace of this rank-2 tensor. Because the trance of a rank-2 tensor is invariant, as established earlier, the divergence of a vector is invariant.

5.18 What is the rank of the divergence of a rank-n ($n > 0$) tensor and why?
The rank is $(n-1)$ because the rank of the original tensor is increased by 1 due to

the action of ∇ and hence it is $(n+1)$. The rank is then reduced by 2 due to the contraction of two indices when taking the trace. Hence the rank of the final tensor is $(n-1)$.

5.19 State, using vector and tensor notations, the mathematical definition of the divergence operation of a vector in a Cartesian coordinate system.

The divergence of a vector $\mathbf{A}$ is given in vector and tensor notations respectively by:

$$\nabla \cdot \mathbf{A} = \frac{\partial A_1}{\partial x_1} + \cdots + \frac{\partial A_n}{\partial x_n}$$
$$\partial_i A_i = A_{i,i}$$

5.20 Discuss in detail the following statement: "The divergence of a vector is a gradient operation followed by a contraction". How this is related to the trace of a rank-2 tensor?

As explained earlier, the divergence of a vector is obtained by taking the gradient of the vector first, which results in a rank-2 tensor, followed by contracting the two indices of this rank-2 tensor to obtain the trace which is the divergence of the vector.

5.21 Write down the mathematical expression of the two forms of the divergence of a rank-2 tensor.
Answer:

$$[\nabla \cdot \mathbf{A}]_i = \partial_j A_{ji} \qquad \text{and} \qquad [\nabla \cdot \mathbf{A}]_j = \partial_i A_{ji}$$

5.22 How many forms do we have for the divergence of a rank-n ($n > 0$) tensor and why? Assume in your answer that the divergence operation can be conducted with respect to any one of the tensor indices.

We have n forms, i.e. one form for each index contracted with the ∇ index.

5.23 Find the divergence of the following vectors $\mathbf{U}$ and $\mathbf{V}$ where x_1, x_2 and x_3 are the Cartesian coordinates:

$$\mathbf{U} = \left(9.3 x_1,\ 6.3 \cos x_2,\ 3.6 x_1 e^{-1.2 x_3}\right) \qquad \mathbf{V} = \left(x_2 \sin x_1,\ 5(x_2)^3,\ 16.3 x_3\right)$$

Answer:

$$\nabla \cdot \mathbf{U} = 9.3 - 6.3 \sin x_2 - 4.32 x_1 e^{-1.2 x_3}$$
$$\nabla \cdot \mathbf{V} = x_2 \cos x_1 + 15(x_2)^2 + 16.3$$

5 APPLICATIONS OF TENSOR NOTATION AND TECHNIQUES

5.24 State, in tensor notation, the mathematical expression for the curl of a vector and of a rank-2 tensor assuming a Cartesian coordinate system.
Answer:
$$[\nabla \times \mathbf{A}]_i = \epsilon_{ijk}\partial_j A_k = \epsilon_{ijk}A_{k,j}$$
$$[\nabla \times \mathbf{A}]_{ij} = \epsilon_{imn}\partial_m A_{nj} = \epsilon_{imn}A_{nj,m}$$

5.25 By using the Index, make a list of terms and notations related to vector and matrix algebra which are used in this chapter.
Some of these terms are: cross product, curl, determinant, divergence, dot product, gradient, Laplacian operator, nabla operator, trace.

5.26 Define, in tensor notation, the Laplacian operator acting on a differentiable scalar field in a Cartesian coordinate system.
Answer:
$$\nabla^2 f = \partial_{ii} f = f_{,ii}$$

5.27 Is the Laplacian a scalar or a vector operator?
It is scalar.

5.28 What is the meaning of the Laplacian operator acting on a differentiable vector field?
The meaning is that the operator acts on the ith component of the vector in a similar manner to its action on a scalar where the result of this action is taken as the ith component of the final vector, that is:
$$\left[\nabla^2 \mathbf{A}\right]_i = \nabla^2 \left[\mathbf{A}\right]_i = \partial_{jj} A_i$$

5.29 What is the rank of a rank-n tensor acted upon by the Laplacian operator?
The rank is n, i.e. the same as the rank of the original tensor.

5.30 Define mathematically the following operators assuming a Cartesian coordinate system:
$$\mathbf{A} \cdot \nabla \qquad\qquad \mathbf{A} \times \nabla$$
where $\mathbf{A}$ is a vector. What is the rank of each one of these operators?

Answer:

$$\mathbf{A} \cdot \nabla = A_i \partial_i$$
$$[\mathbf{A} \times \nabla]_i = \epsilon_{ijk} A_j \partial_k$$

The rank of the first is 0 (i.e. it is scalar operator) and the rank of the second is 1 (i.e. it is vector operator).

5.31 Make a general statement about how differentiation of tensors affects their rank discussing in detail from this perspective the gradient and divergence operations.
The differentiation of a tensor by the above nabla-based operations generally increases its rank by one, by introducing an extra free covariant index on the original tensor, unless it implies a contraction in which case it reduces the rank by one due to the consumption of the differentiation index with one of the original tensor indices by the contraction operation. Therefore, the gradient of a scalar is a vector and the gradient of a vector is a rank-2 tensor since the gradient is a nabla-based differential operation without contraction. On the other hand, the divergence of a vector is a scalar and the divergence of a rank-2 tensor is a vector because the divergence is a nabla-based differential operation (considering the outer product of nabla and tensor) with contraction. Regarding the curl operation, it is not an ordinary differential operation since it is based on taking the cross product of the nabla operator with the tensor and hence we are not differentiating the tensor in an ordinary manner. Accordingly, the curl operation does not change the rank. Regarding the Laplacian, it is not a differential operation by the nabla operator because the Laplacian is the divergence of the gradient (or the divergence of nabla) and hence it is a scalar operator, unlike nabla which is a vector operator. Accordingly, the Laplacian operator does not change the rank of the tensor since it has no free differentiation index to add to the original tensor.

5.32 State the mathematical expressions for the following operators and operations assuming a cylindrical coordinate system: nabla operator, Laplacian operator, gradient of a scalar, divergence of a vector, and curl of a vector.
Answer:

$$\nabla = \mathbf{e}_\rho \partial_\rho + \mathbf{e}_\phi \frac{1}{\rho} \partial_\phi + \mathbf{e}_z \partial_z$$

5 APPLICATIONS OF TENSOR NOTATION AND TECHNIQUES

$$\nabla^2 = \partial_{\rho\rho} + \frac{1}{\rho}\partial_\rho + \frac{1}{\rho^2}\partial_{\phi\phi} + \partial_{zz}$$

$$\nabla f = \mathbf{e}_\rho \frac{\partial f}{\partial \rho} + \mathbf{e}_\phi \frac{1}{\rho}\frac{\partial f}{\partial \phi} + \mathbf{e}_z \frac{\partial f}{\partial z}$$

$$\nabla \cdot \mathbf{A} = \frac{1}{\rho}\left[\frac{\partial(\rho A_\rho)}{\partial \rho} + \frac{\partial A_\phi}{\partial \phi} + \frac{\partial(\rho A_z)}{\partial z}\right]$$

$$\nabla \times \mathbf{A} = \frac{1}{\rho}\begin{vmatrix} \mathbf{e}_\rho & \rho\mathbf{e}_\phi & \mathbf{e}_z \\ \frac{\partial}{\partial \rho} & \frac{\partial}{\partial \phi} & \frac{\partial}{\partial z} \\ A_\rho & \rho A_\phi & A_z \end{vmatrix}$$

where the symbols are as defined in the text.

5.33 Explain how the expressions for the operators and operations in the previous exercise can be obtained for the plane polar coordinate system from the expressions of the cylindrical system.

For the 2D plane polar coordinate systems, the expressions of these operators and operations can be obtained by dropping the z components or terms from the cylindrical form of these operators and operations, as given in the previous exercise. This is inline with the fact that the plane polar coordinate system is a cylindrical coordinate system without z dimension.

5.34 State the mathematical expressions for the following operators and operations assuming a spherical coordinate system: nabla operator, Laplacian operator, gradient of a scalar, divergence of a vector, and curl of a vector.

Answer:

$$\nabla = \mathbf{e}_r \partial_r + \mathbf{e}_\theta \frac{1}{r}\partial_\theta + \mathbf{e}_\phi \frac{1}{r\sin\theta}\partial_\phi$$

$$\nabla^2 = \partial_{rr} + \frac{2}{r}\partial_r + \frac{1}{r^2}\partial_{\theta\theta} + \frac{\cos\theta}{r^2\sin\theta}\partial_\theta + \frac{1}{r^2\sin^2\theta}\partial_{\phi\phi}$$

$$\nabla f = \mathbf{e}_r \frac{\partial f}{\partial r} + \mathbf{e}_\theta \frac{1}{r}\frac{\partial f}{\partial \theta} + \mathbf{e}_\phi \frac{1}{r\sin\theta}\frac{\partial f}{\partial \phi}$$

$$\nabla \cdot \mathbf{A} = \frac{1}{r^2\sin\theta}\left[\sin\theta\frac{\partial(r^2 A_r)}{\partial r} + r\frac{\partial(\sin\theta A_\theta)}{\partial \theta} + r\frac{\partial A_\phi}{\partial \phi}\right]$$

$$\nabla \times \mathbf{A} = \frac{1}{r^2\sin\theta}\begin{vmatrix} \mathbf{e}_r & r\mathbf{e}_\theta & r\sin\theta\mathbf{e}_\phi \\ \frac{\partial}{\partial r} & \frac{\partial}{\partial \theta} & \frac{\partial}{\partial \phi} \\ A_r & rA_\theta & r\sin\theta A_\phi \end{vmatrix}$$

where the symbols are as defined in the text.

5.35 Repeat the previous exercise for the general orthogonal coordinate system.

Assuming a 3D space, we have:

$$\nabla = \frac{\mathbf{u}_1}{h_1}\frac{\partial}{\partial u^1} + \frac{\mathbf{u}_2}{h_2}\frac{\partial}{\partial u^2} + \frac{\mathbf{u}_3}{h_3}\frac{\partial}{\partial u^3}$$

$$\nabla^2 = \frac{1}{h_1 h_2 h_3}\left[\frac{\partial}{\partial u^1}\left(\frac{h_2 h_3}{h_1}\frac{\partial}{\partial u^1}\right) + \frac{\partial}{\partial u^2}\left(\frac{h_1 h_3}{h_2}\frac{\partial}{\partial u^2}\right) + \frac{\partial}{\partial u^3}\left(\frac{h_1 h_2}{h_3}\frac{\partial}{\partial u^3}\right)\right]$$

$$\nabla f = \frac{\mathbf{u}_1}{h_1}\frac{\partial f}{\partial u^1} + \frac{\mathbf{u}_2}{h_2}\frac{\partial f}{\partial u^2} + \frac{\mathbf{u}_3}{h_3}\frac{\partial f}{\partial u^3}$$

$$\nabla \cdot \mathbf{A} = \frac{1}{h_1 h_2 h_3}\left[\frac{\partial}{\partial u^1}(h_2 h_3 A_1) + \frac{\partial}{\partial u^2}(h_1 h_3 A_2) + \frac{\partial}{\partial u^3}(h_1 h_2 A_3)\right]$$

$$\nabla \times \mathbf{A} = \frac{1}{h_1 h_2 h_3}\begin{vmatrix} h_1 \mathbf{u}_1 & h_2 \mathbf{u}_2 & h_3 \mathbf{u}_3 \\ \frac{\partial}{\partial u^1} & \frac{\partial}{\partial u^2} & \frac{\partial}{\partial u^3} \\ h_1 A_1 & h_2 A_2 & h_3 A_3 \end{vmatrix}$$

where the symbols are as defined in the text.

5.36 Express, in tensor notation, the mathematical condition for a vector field to be solenoidal.

Answer:
$$\partial_i A_i = 0$$

5.37 Express, in tensor notation, the mathematical condition for a vector field to be irrotational.

Answer:
$$\epsilon_{ijk}\partial_j A_k = 0$$

5.38 Express, in tensor notation, the divergence theorem for a differentiable vector field explaining all the symbols involved. Repeat the exercise for a differentiable tensor field of an arbitrary rank (> 0).

The divergence theorem for a differentiable vector field A_i is given in tensor notation by:

$$\int_V \partial_i A_i \, d\tau = \int_S A_i n_i \, d\sigma$$

where ∂_i is the nabla operator, V is a bounded region in an nD space enclosed by a generalized surface S, $d\tau$ and $d\sigma$ are generalized volume and area elements respectively, n_i is the ith component of the unit vector normal to the surface, and the index i ranges over $1, \ldots, n$.

5 APPLICATIONS OF TENSOR NOTATION AND TECHNIQUES

The divergence theorem for a differentiable non-scalar tensor field of arbitrary rank $A_{i \cdots m}$ is given in tensor notation by:

$$\int_V \partial_k A_{i \ldots k \ldots m} d\tau = \int_S A_{i \ldots k \ldots m} n_k d\sigma$$

where the symbols are as explained in the previous part.

5.39 Express, in tensor notation, Stokes theorem for a differentiable vector field explaining all the symbols involved. Repeat the exercise for a differentiable tensor field of an arbitrary rank (> 0).

Stokes theorem for a differentiable vector field A_i is given in tensor notation by:

$$\int_S \epsilon_{ijk} \partial_j A_k n_i d\sigma = \int_C A_i dx_i$$

where ϵ_{ijk} is the rank-3 permutation tensor, C stands for the perimeter of the surface S, and dx_i is the ith component of an infinitesimal vector element tangent to the perimeter while the other symbols are as defined in the previous exercise.

Stokes theorem for a differentiable non-scalar tensor field of arbitrary rank $A_{l \cdots n}$ is given in tensor notation by:

$$\int_S \epsilon_{ijk} \partial_j A_{l \ldots k \ldots n} n_i d\sigma = \int_C A_{l \ldots k \ldots n} dx_k$$

where the symbols are as explained in the previous part.

5.40 Express the following identities in tensor notation:

$$\nabla \cdot \mathbf{r} = n$$
$$\nabla (\mathbf{a} \cdot \mathbf{r}) = \mathbf{a}$$
$$\nabla \cdot (\nabla \times \mathbf{A}) = 0$$
$$\nabla (fh) = f\nabla h + h\nabla f$$
$$\nabla \times (f\mathbf{A}) = f\nabla \times \mathbf{A} + \nabla f \times \mathbf{A}$$
$$\mathbf{A} \times (\mathbf{B} \times \mathbf{C}) = \mathbf{B}(\mathbf{A} \cdot \mathbf{C}) - \mathbf{C}(\mathbf{A} \cdot \mathbf{B})$$
$$\nabla \times (\nabla \times \mathbf{A}) = \nabla (\nabla \cdot \mathbf{A}) - \nabla^2 \mathbf{A}$$
$$\nabla \cdot (\mathbf{A} \times \mathbf{B}) = \mathbf{B} \cdot (\nabla \times \mathbf{A}) - \mathbf{A} \cdot (\nabla \times \mathbf{B})$$

$$(\mathbf{A} \times \mathbf{B}) \cdot (\mathbf{C} \times \mathbf{D}) = \begin{vmatrix} \mathbf{A} \cdot \mathbf{C} & \mathbf{A} \cdot \mathbf{D} \\ \mathbf{B} \cdot \mathbf{C} & \mathbf{B} \cdot \mathbf{D} \end{vmatrix}$$

These identities are given in tensor notation by:

$$\partial_i x_i = n$$
$$\partial_i (a_j x_j) = a_i$$
$$\epsilon_{ijk} \partial_i \partial_j A_k = 0$$
$$\partial_i (fh) = f \partial_i h + h \partial_i f$$
$$\epsilon_{ijk} \partial_j (f A_k) = f \epsilon_{ijk} \partial_j A_k + \epsilon_{ijk} (\partial_j f) A_k$$
$$\epsilon_{ijk} A_j \epsilon_{klm} B_l C_m = B_i (A_m C_m) - C_i (A_l B_l)$$
$$\epsilon_{ijk} \epsilon_{klm} \partial_j \partial_l A_m = \partial_i (\partial_m A_m) - \partial_{ll} A_i$$
$$\partial_i (\epsilon_{ijk} A_j B_k) = B_k (\epsilon_{kij} \partial_i A_j) - A_j (\epsilon_{jik} \partial_i B_k)$$
$$\epsilon_{ijk} A_j B_k \epsilon_{ilm} C_l D_m = (A_l C_l)(B_m D_m) - (A_m D_m)(B_l C_l)$$

5.41 Prove the following identities using the language and techniques of tensor calculus:

$$\nabla \times \mathbf{r} = \mathbf{0}$$
$$\nabla \cdot (\nabla f) = \nabla^2 f$$
$$\nabla \times (\nabla f) = \mathbf{0}$$
$$\nabla \cdot (f \mathbf{A}) = f \nabla \cdot \mathbf{A} + \mathbf{A} \cdot \nabla f$$
$$\mathbf{A} \cdot (\mathbf{B} \times \mathbf{C}) = \mathbf{C} \cdot (\mathbf{A} \times \mathbf{B}) = \mathbf{B} \cdot (\mathbf{C} \times \mathbf{A})$$
$$\mathbf{A} \times (\nabla \times \mathbf{B}) = (\nabla \mathbf{B}) \cdot \mathbf{A} - \mathbf{A} \cdot \nabla \mathbf{B}$$
$$\nabla (\mathbf{A} \cdot \mathbf{B}) = \mathbf{A} \times (\nabla \times \mathbf{B}) + \mathbf{B} \times (\nabla \times \mathbf{A}) + (\mathbf{A} \cdot \nabla) \mathbf{B} + (\mathbf{B} \cdot \nabla) \mathbf{A}$$
$$\nabla \times (\mathbf{A} \times \mathbf{B}) = (\mathbf{B} \cdot \nabla) \mathbf{A} + (\nabla \cdot \mathbf{B}) \mathbf{A} - (\nabla \cdot \mathbf{A}) \mathbf{B} - (\mathbf{A} \cdot \nabla) \mathbf{B}$$
$$(\mathbf{A} \times \mathbf{B}) \times (\mathbf{C} \times \mathbf{D}) = [\mathbf{D} \cdot (\mathbf{A} \times \mathbf{B})] \mathbf{C} - [\mathbf{C} \cdot (\mathbf{A} \times \mathbf{B})] \mathbf{D}$$

The proofs of these identities are given in the text.

Chapter 6

Metric Tensor

6.1 Describe in details, using mathematical tensor language when necessary, the metric tensor discussing its rank, purpose, designations, variance types, symmetry, its role in the definition of distance, and its relation to the covariant and contravariant basis vectors.

The metric tensor is a rank-2 symmetric absolute non-singular tensor. This tensor is one of the most important special tensors in tensor calculus and its applications. As a tensor, the metric has significance regardless of any coordinate system although it requires a coordinate system to be represented in a specific form. This tensor is used, for instance, to raise and lower indices and thus facilitate the transformation between the covariant and contravariant types. One of the main objectives of the metric tensor is to generalize the concept of distance to general coordinate systems and hence maintain the invariance of distance in different coordinate systems. In orthonormal Cartesian coordinate systems the distance element squared, $(ds)^2$, between two infinitesimally neighboring points in space, one with coordinates x_i and the other with coordinates $x_i + dx_i$, is given by:

$$(ds)^2 = dx_i dx_i = \delta_{ij} dx_i dx_j$$

This definition of distance is the key to introducing the metric tensor which, for a general coordinate system, is defined by:

$$(ds)^2 = g_{ij} dx^i dx^j$$

The metric tensor can be of covariant form g_{ij}, or contravariant form g^{ij}, or mixed form g^i_j where the mixed form is same as the unity tensor δ^i_j. The covariant and contravariant forms of the metric tensor are inverses of each other and hence we have:

$$g^{ik} g_{kj} = \delta^i_j \qquad\qquad g_{ik} g^{kj} = \delta^j_i$$

The components of the metric tensor in its different forms are related to the basis vectors of the coordinate system by:

$$g_{ij} = \mathbf{E}_i \cdot \mathbf{E}_j$$
$$g^{ij} = \mathbf{E}^i \cdot \mathbf{E}^j$$
$$g^i_j = \mathbf{E}^i \cdot \mathbf{E}_j$$

6.2 What is the relation between the covariant and contravariant types of the metric tensor? Express this relation mathematically. Also define mathematically the mixed type metric tensor.

They are inverses of each other, that is:

$$g^{ik} g_{kj} = \delta^i_j \qquad\qquad g_{ik} g^{kj} = \delta^j_i$$

The mixed type metric tensor is the same as the unity tensor δ^i_j, that is:

$$g^i_j = \mathbf{E}^i \cdot \mathbf{E}_j = \delta^i_j$$

6.3 Correct, if necessary, the following equations:

$$g^i_j = \delta^i_i \qquad g^{ij} = \mathbf{E}^i \cdot \mathbf{E}_j \qquad (ds) = g_{ij} dx^i dx^j \qquad g_{ij} = \mathbf{E}_i \cdot \mathbf{E}_j \qquad \mathbf{E}^i \cdot \mathbf{E}_j = \delta^j_i$$

The fourth equation is correct. The correct form of the other equations is:

$$g^i_j = \delta^i_j \qquad g^{ij} = \mathbf{E}^i \cdot \mathbf{E}^j \qquad (ds)^2 = g_{ij} dx^i dx^j \qquad \mathbf{E}^i \cdot \mathbf{E}_j = \delta^i_j$$

6.4 What "flat metric" means? Give an example of a coordinate system with a flat metric.
"Flat metric" means a metric tensor of a space that can have a coordinate system with a diagonal metric tensor whose all diagonal elements are ± 1. An example of a coordinate system of a space with a flat metric is the orthonormal Cartesian system of a Euclidean space whose metric is a diagonal tensor with all its diagonal elements being $+1$. Any metric of such a space is a flat metric.

6.5 Describe the index-shifting (raising/lowering) operators and their relation to the metric tensor. How these operators facilitate the transformation between the covariant, contravariant and mixed types of a given tensor?

The contravariant form of the metric tensor can be used to raise a lower index and the covariant form of the metric tensor can be used to lower an upper index and hence they are described as index-shifting (i.e. raising or lowering) operators. These operators act by contracting one of their indices with an index of the opposite variance type of another tensor in their own term, hence replacing that contracted index in the other tensor by their other index with changing its position. For example, we can raise the index of a covariant vector A_j by the following operation:

$$A^i = g^{ij} A_j$$

Similarly, we can lower the index of a contravariant vector B^j by the following operation:

$$B_i = g_{ij} B^j$$

Thus, the covariant and contravariant forms of the metric tensor allow the shift of indices from one variance type to the opposite variance type and hence they facilitate the transformation between the covariant, contravariant and mixed types of a given tensor.

6.6 Find from the Index all the names and labels of the metric tensor in its covariant, contravariant and mixed types.

Associate metric tensor, conjugate metric tensor, contravariant metric tensor, covariant metric tensor, fundamental tensor, index lowering operator, index raising operator, index shifting operator, mixed metric tensor, reciprocal metric tensor.
Labels like: "identity tensor, index replacement operator, Kronecker δ, unit tensor, unity tensor" can also apply to the metric tensor in its mixed form.

6.7 What is wrong with the following equations?

$$C_i = g^{ij} C_j \qquad\qquad D_i = g^{ij} D^j \qquad\qquad A^i = \delta_{ij} A_j$$

Make the necessary corrections considering all the possibilities in each case.
The metric tensor in these equations is not used properly as an index raising, index lowering and index replacement operator. The correct form of these equations is:

$$C^i = g^{ij} C_j \qquad\qquad D_i = g_{ij} D^j \qquad\qquad A^i = \delta^i_j A^j$$

There are other possibilities like $D^i = g^{ij}D_j$ for the second equation and $A^i = g^{ij}A_j$ and $A_i = \delta_i^j A_j$ for the third equation. Also, the third equation can be correct if the coordinate system is orthonormal Cartesian although it will be more appropriate in this case to be given in covariant form, i.e. $A_i = \delta_{ij}A_j$.

6.8 Is it necessary to keep the order of the indices which are shifted by the index-shifting operators and why?
Yes, because two tensors with the same indicial structure but with different indicial order are not equal in general.

6.9 How and why dots may be inserted to avoid confusion about the order of the indices following an index-shifting operation?
A dot may be inserted in the original position of the shifted index to record the position of the shifted index prior to its shift. As well as removing any ambiguity about the order of indices, the dots enable the reversal of shifting if needed in the future since the original position of the shifted index is documented.

6.10 Express, mathematically, the fact that the contravariant and covariant metric tensors are inverses of each other.
This fact can be expressed as:

$$[g_{ij}] = [g^{ij}]^{-1} \qquad\qquad [g^{ij}] = [g_{ij}]^{-1}$$

or as:

$$g^{ik}g_{kj} = \delta^i_j \qquad\qquad g_{ik}g^{kj} = \delta_i^j$$

6.11 Collect from the Index a list of operators and operations which have particular links to the metric tensor.
Index lowering operator, index raising operator, index replacement operator, index shifting operator, substitution operator.

6.12 Correct, if necessary, the following statement: "The term metric tensor is usually used to label the covariant form of the metric, while the contravariant form of the metric is called the conjugate or associate or reciprocal metric tensor".
This statement is correct.

6.13 Write, in matrix form, the covariant and contravariant types of the metric tensor of the Cartesian, cylindrical and spherical coordinate systems of a 3D flat space.

6 METRIC TENSOR

Cartesian:
$$[g_{ij}] = [g^{ij}] = \begin{bmatrix} 1 & 0 & 0 \\ 0 & 1 & 0 \\ 0 & 0 & 1 \end{bmatrix}$$

Cylindrical:
$$[g_{ij}] = \begin{bmatrix} 1 & 0 & 0 \\ 0 & \rho^2 & 0 \\ 0 & 0 & 1 \end{bmatrix} \qquad [g^{ij}] = \begin{bmatrix} 1 & 0 & 0 \\ 0 & \frac{1}{\rho^2} & 0 \\ 0 & 0 & 1 \end{bmatrix}$$

Spherical:
$$[g_{ij}] = \begin{bmatrix} 1 & 0 & 0 \\ 0 & r^2 & 0 \\ 0 & 0 & r^2 \sin^2 \theta \end{bmatrix} \qquad [g^{ij}] = \begin{bmatrix} 1 & 0 & 0 \\ 0 & \frac{1}{r^2} & 0 \\ 0 & 0 & \frac{1}{r^2 \sin^2 \theta} \end{bmatrix}$$

6.14 Regarding the previous question, what do you notice about the corresponding diagonal elements of the covariant and contravariant types of the metric tensor in these systems? Does this relate to the fact that these types are inverses of each other?

The corresponding diagonal elements of the covariant and contravariant types of the metric tensor in these systems are reciprocals.

Yes, because the inverse of an invertible diagonal matrix is a diagonal matrix obtained by taking the reciprocal of the corresponding diagonal elements of the original matrix.

Chapter 7
Covariant Differentiation

7.1 Is the ordinary derivative of a tensor necessarily a tensor or not? Can the ordinary derivative of a tensor be a tensor? If so, give an example.
It is not necessarily a tensor although it can be.
Yes. The ordinary derivative of a rank-0 tensor (scalar) is tensor. Also, the ordinary derivatives of non-scalar tensors in rectilinear coordinate systems (e.g. orthonormal Cartesian) are tensors.

7.2 Explain the purpose of the covariant derivative and how it is related to the invariance property of tensors.
The purpose of covariant derivative is to generalize the definition of ordinary partial derivative so that when this type of derivative (i.e. covariant derivative) is applied to a tensor it will necessarily produce a tensor and hence it satisfies the principle of tensor invariance. This is achieved by extending the definition of partial derivative to include the basis vectors as well as the tensor components.

7.3 Is the covariant derivative of a tensor necessarily a tensor? If so, what is the rank of the covariant derivative of a rank-n tensor?
Yes. The rank is $(n+1)$.

7.4 How is the Christoffel symbol of the second kind symbolized? Describe the arrangement of its indices.
The Christoffel symbol of the second kind is symbolized in the literature in a number of forms. The form that we use in the text is a pair of curly brackets between which the three indices of the Christoffel symbol are contained where the first two indices are located in the bottom while the third index is located in the top, e.g. $\{{}^{k}_{ij}\}$.

7.5 State the mathematical definition of the Christoffel symbol of the second kind in terms of the metric tensor defining all the symbols involved.
The Christoffel symbol of the second kind $\{{}^{k}_{ij}\}$ is defined by:

$$\{{}^{k}_{ij}\} = \frac{g^{kl}}{2}\left(\frac{\partial g_{il}}{\partial x^j} + \frac{\partial g_{jl}}{\partial x^i} - \frac{\partial g_{ij}}{\partial x^l}\right)$$

where the indexed g represent the components of the metric tensor in its contravariant (g^{kl}) and covariant (g_{il}, g_{jl}, g_{ij}) forms with implied summation over l, ∂ is the partial differential symbol and the indexed x symbolize the coordinates of a general coordinate system.

7.6 The Christoffel symbols of the second kind are symmetric in which of their indices?
They are symmetric in their two lower indices, that is: $\{{}^k_{ij}\} = \{{}^k_{ji}\}$.

7.7 Why the Christoffel symbols of the second kind are identically zero in the Cartesian coordinate systems? Use in your explanation the mathematical definition of the Christoffel symbols.
Because the components of the metric tensor in Cartesian coordinate systems are constant. Hence, all the partial derivatives in the definition of the Christoffel symbols of the second kind, as given above, will vanish resulting in having identically zero Christoffel symbols in these systems.

7.8 Give the Christoffel symbols of the second kind for the cylindrical and spherical coordinate systems explaining the meaning of the indices used.
In cylindrical coordinate systems, identified by the coordinates (ρ, ϕ, z), the Christoffel symbols of the second kind are zero for all values of the indices except:

$$\{{}^1_{22}\} = -\rho$$
$$\{{}^2_{12}\} = \{{}^2_{21}\} = \frac{1}{\rho}$$

where the indices $(1, 2)$ stand for (ρ, ϕ).
For spherical coordinate systems, identified by the coordinates (r, θ, ϕ), the Christoffel symbols of the second kind are zero for all values of the indices except:

$$\{{}^1_{22}\} = -r$$
$$\{{}^1_{33}\} = -r\sin^2\theta$$
$$\{{}^2_{12}\} = \{{}^2_{21}\} = \frac{1}{r}$$
$$\{{}^2_{33}\} = -\sin\theta\cos\theta$$
$$\{{}^3_{13}\} = \{{}^3_{31}\} = \frac{1}{r}$$
$$\{{}^3_{23}\} = \{{}^3_{32}\} = \cot\theta$$

7 COVARIANT DIFFERENTIATION

where the indices $(1,2,3)$ stand for (r,θ,ϕ).

7.9 From the Index, collect five terms used in the definition and description of the Christoffel symbols of the second kind.
Contravariant metric tensor, covariant metric tensor, differential operator, partial differentiation, summation convention.

7.10 What is the meaning, within the context of tensor differentiation, of the comma "," and semicolon ";" when used as subscripts preceding a tensor index?
The subscript comma means ordinary partial differentiation of the tensor with respect to the coordinate indexed by the index that follows the comma. The subscript semicolon means covariant differentiation of the tensor with respect to the coordinate indexed by the index that follows the semicolon. Hence, $A_{i,j}$ means partial derivative of tensor A_i with respect to x^j while $A_{i;j}$ means covariant derivative of tensor A_i with respect to x^j.

7.11 Why the covariant derivative of a differentiable scalar is the same as the ordinary partial derivative and how does this relate to the basis vectors of coordinate systems?
Because a scalar does not refer to a basis vector set, hence its covariant derivative applies only to the scalar quantity which is the same as the ordinary partial derivative. Alternatively, a scalar does not have a free index, hence it has no Christoffel symbol terms, so what remains of the covariant derivative terms is the ordinary partial derivative term only.

7.12 Differentiate the following tensors covariantly:

$$A^s \qquad B_t \qquad C_i^j \qquad D_{pq} \qquad E^{mn} \qquad A_{ij...k}^{lm...p}$$

Answer:

$$
\begin{aligned}
A^s_{;i} &= \partial_i A^s + \{{}^{s}_{ki}\} A^k \\
B_{t;i} &= \partial_i B_t - \{{}^{k}_{ti}\} B_k \\
C^j_{i;q} &= \partial_q C^j_i + \{{}^{j}_{lq}\} C^l_i - \{{}^{l}_{iq}\} C^j_l \\
D_{pq;i} &= \partial_i D_{pq} - \{{}^{l}_{pi}\} D_{lq} - \{{}^{l}_{qi}\} D_{pl} \\
E^{mn}_{;i} &= \partial_i E^{mn} + \{{}^{m}_{li}\} E^{ln} + \{{}^{n}_{li}\} E^{ml} \\
A^{lm...p}_{ij...k;q} &= \partial_q A^{lm...p}_{ij...k} + \{{}^{l}_{aq}\} A^{am...p}_{ij...k} + \{{}^{m}_{aq}\} A^{la...p}_{ij...k} + \cdots + \{{}^{p}_{aq}\} A^{lm...a}_{ij...k}
\end{aligned}
$$

7 COVARIANT DIFFERENTIATION

$$-\left\{{a \atop iq}\right\} A^{lm...p}_{aj...k} - \left\{{a \atop jq}\right\} A^{lm...p}_{ia...k} - \cdots - \left\{{a \atop kq}\right\} A^{lm...p}_{ij...a}$$

7.13 Explain the mathematical pattern followed in the operation of covariant differentiation of tensors. Does this pattern also apply to rank-0 tensors?

The pattern of covariant differentiation is that it starts with an ordinary partial derivative term of the component of the given tensor. Then for each free index of the tensor an extra Christoffel symbol term is added where the Christoffel symbol term satisfies the following properties:

(a) The term is positive for contravariant indices and negative for covariant indices.
(b) One of the lower indices of the Christoffel symbol is the differentiation index.
(c) The differentiated index of the tensor in the concerned Christoffel symbol term is contracted with one of the indices of the Christoffel symbol using a new label and hence they are opposite in their variance type.
(d) The label of the differentiated index is transferred from the tensor to the Christoffel symbol keeping its position as lower or upper.
(e) All the other indices of the tensor in the concerned Christoffel symbol term keep their labels, position and order.

This pattern also applies to rank-0 tensors since scalars have no free index and hence there is no Christoffel symbol term in their covariant derivative so what remains is the ordinary partial derivative term which is equivalent to the previously established fact that the covariant derivative of a scalar is the same as its ordinary partial derivative.

7.14 The covariant derivative in Cartesian coordinate systems is the same as the ordinary partial derivative for all tensor ranks. Explain why.

Because all the Christoffel symbols vanish identically in Cartesian systems, so all the Christoffel symbol terms in the covariant derivative expression are zero. Hence, only the ordinary partial derivative term in the covariant derivative expression remains which means that the covariant derivative in Cartesian systems is the same as the ordinary partial derivative.

7.15 What is the covariant derivative of the covariant and contravariant forms of the metric tensor for an arbitrary type of coordinate system? How is this related to the fact that the covariant derivative operator bypasses the index-shifting operator?

The covariant derivative of all forms of the metric tensor is identically zero in any coordinate system, so the metric tensor behaves like a constant with respect to covariant differentiation. Accordingly, when covariant differentiation is applied on a

7 COVARIANT DIFFERENTIATION

tensor product that involves the metric tensor acting as an index-shifting operator, the covariant derivative operator can bypass the shifting operator, i.e. the covariant derivative operator and the shifting operator commute.

7.16 Which rules of ordinary differentiation apply equally to covariant differentiation and which do not? Make mathematical statements about all these rules with sufficient explanation of the symbols and operations involved.

The sum and product rules of ordinary differentiation apply, but commutativity of ordinary differential operators does not apply, that is:

$$\begin{aligned}
\partial_{;i}(a\mathbf{A} \pm b\mathbf{B}) &= a\,\partial_{;i}\mathbf{A} \pm b\,\partial_{;i}\mathbf{B} \\
\partial_{;i}(\mathbf{AB}) &= (\partial_{;i}\mathbf{A})\mathbf{B} + \mathbf{A}(\partial_{;i}\mathbf{B}) \\
\partial_{;i}\partial_{;j} &\neq \partial_{;j}\partial_{;i}
\end{aligned}$$

where a and b are scalar constants and $\mathbf{A}$ and $\mathbf{B}$ are differentiable tensor fields. Also, unlike ordinary differentiation, the order of the tensors should be observed in the product rule since tensor multiplication, unlike ordinary multiplication, is not commutative. The product rule also applies to inner product of tensors as well as outer product.

7.17 Make corrections, where necessary, in the following equations explaining in each case why the equation should or should not be amended:

$$\begin{aligned}
(\mathbf{C} \pm \mathbf{D})_{;i} &= \partial_{;i}\mathbf{C} \pm \partial_{;i}\mathbf{D} \\
\partial_{;i}(\mathbf{AB}) &= \mathbf{B}(\partial_{;i}\mathbf{A}) + \mathbf{A}\partial_{;i}\mathbf{B} \\
\left(g_{ij}A^j\right)_{;m} &= g_{ij}\left(A^j\right)_{;m} \\
\partial_{;i}\partial_{;j} &= \partial_{;j}\partial_{;i} \\
\partial_i\partial_j &= \partial_j\partial_i
\end{aligned}$$

First equation: correct by the sum rule with the use of various notations.
Second equation should be:

$$\partial_{;i}(\mathbf{AB}) = (\partial_{;i}\mathbf{A})\mathbf{B} + \mathbf{A}\partial_{;i}\mathbf{B}$$

because tensor multiplication is not commutative and hence the order should be kept.
Third equation: correct, because the metric tensor commutes with the covariant

7 COVARIANT DIFFERENTIATION

differential operator since the metric tensor behaves like a constant with respect to covariant differentiation.

Fourth equation: incorrect, because covariant differential operators with respect to different indices do not commute. Hence, it should be:

$$\partial_{;i}\partial_{;j} \neq \partial_{;j}\partial_{;i}$$

Fifth equation: correct, because ordinary partial differential operators with respect to different indices do commute, assuming the C^2 continuity condition.

7.18 How do you define the second and higher order covariant derivatives of tensors? Do these derivatives follow the same rules as the ordinary partial derivatives of the same order in the case of different differentiation indices?

Second and higher order covariant derivatives of tensors are defined as derivatives of derivatives, like their ordinary counterparts.

No, because the order of differentiation should be respected in the case of covariant differentiation but it is irrelevant in the case of ordinary partial differentiation.

7.19 From the Index, find all the terms that refer to symbols used in the notation of ordinary partial derivatives and covariant derivatives.

Comma notation, semicolon notation.

Printed in Poland
by Amazon Fulfillment
Poland Sp. z o.o., Wrocław